WRITTEN IN STONE

A GEOLOGICAL HISTORY OF THE NORTHEASTERN UNITED STATES

Chet Raymo &
Maureen E. Raymo

BLACK · DOME
Black Dome Press Corp.
1011 Route 296
Hensonville, New York 12439
Tel: (518) 734-6357
Fax: (518) 734-5802
www.blackdomepress.com

For Dan and Vicky

Published by
Black Dome Press Corp.
1011 Route 296
Hensonville, New York 12439
www.blackdomepress.com
Tel: (518) 734-6357 Fax: (518) 734-5802

Second Edition

Library of Congress Cataloging-in-Publication Data:
Raymo, Chet.
Written in stone : a geological history of the northeastern United States / Chet Raymo & Maureen E. Raymo.--2nd ed.
p. cm.
Includes bibliographical references and index.
ISBN 1-883789-27-3 (trade paper)
1. Geology--Northeastern States. 2. Paleontology--Northeastern States. I. Raymo, Maureen E. II. Title.

QE78.3 .R39 2001
557.4--dc21

2001025111

Cover photograph by Thomas Teich
Map cut-away and time tables by Paula K. Hopson
Illustrations by Gil Fahey
Cover design by Carol Clement, Artemisia, Inc.
Printed in the USA

10 9 8 7 6 5 4

Preface to the Second Edition

Many exciting discoveries have been made in the Earth Sciences in the eleven years since this book was first published. Scientists have found evidence (albeit controversial) for past life on Mars, advances in dating have revolutionized our understanding of the rate of evolution early in Earth's history, and evidence for catastrophic flooding of the Black Sea at the end of the last ice age may point to the origin of the Noah's Flood myth. A few discoveries in particular are relevant to the geologic history of the Northeast discussed in our book. Rather than awkwardly rewriting selected passages we mention those advances here (and thank our new publisher for giving us this opportunity!) One of the most exciting discoveries of the nineties was the identification of the crater believed to be associated with the asteroid impact that caused the extinction of the dinosaurs 65 million years ago (Chapter 8). The 120-mile-diameter Chicxulub crater, one of the largest craters of the last 4 billion years, was discovered at the tip of the Yucatan Peninsula in Mexico. Dating of crater rocks show it to be the same age as the global extinction horizon identified elsewhere around the globe.

Another remarkable discovery has been made by geologists studying the recent climate history of North America and Europe. The transition between cold glacial climates and warm interglacial climates that occurred about 12,000 years ago (Chapter 9) appears to have happened over decades or centuries rather than the few thousands of years that had previously been allotted for this dramatic climate change. Apparently, the Earth's climate is able to "flip" between different modes, an astonishing discovery that lends added weight to society's concern about anthropogenic, or human-induced, climate changes at the dawn of the twenty-first century. The last decade has seen the growth of a nearly unanimous consensus amongst climate scientists that the well-documented global warming of the twentieth century is partially due to human activities, in particular the pollution of the atmosphere with greenhouse gases such as carbon dioxide. Few scientists anticipate that this trend will alter over the course of the next century—unless humankind collectively adopts a more caring attitude toward Mother Earth.

Lastly, we made changes to ages quoted in the book to reflect the most recent advances in the geologic timescale. The ages used here derive from the 1999 Geologic Time Scale of The Geological Society of America, the details of which can be found on their web site. We rounded off all ages older than 100 million years to the nearest 5 million.

MER and CTR
Boston, 2001

Contents

About the Authors

Chet Raymo is a professor of physics and astronomy at Stonehill College in Massachusetts. For thirty-five years he has explored the relationships between science, nature and the humanities as a teacher, writer, and columnist for the *Boston Globe*. He is the author of *365 Starry Nights, Honey from Stone*, and *Natural Prayers*, as well as other books.

Maureen E. Raymo, Chet's daughter, is a professor of Earth Science at Boston University where she teaches and conducts research. She has authored numerous scientific publications on the topic of Earth's climate and how it has changed in the past.

Introduction

Imagine yourself standing on the bank of the Connecticut River near Hartford, Connecticut. The river flows placidly through green-mantled hills to the sea. The air is scented by flowering plants, and grasses weave and wave on the shore. Deer graze at the water's edge; birds sing. In the distance are the skyscrapers of the city, the airport and highways, and the sprawling suburbs, all energized by human activity. Now imagine yourself standing on the same piece of land 200 million years ago. Tall mountains cast long shadows across a bleak desert valley. Volcanos spew noxious gases into the air, and rivers of lava burst from the rent earth. The land shakes. Dinosaurs cease their feeding and fearfully test the air. More remarkably, the piece of crust on which you stand is much nearer the Earth's equator, 1,000 miles south of where it is today.

How did things change so much? In this book we explain, step by step, how extraordinary processes, which persist to this day, have continuously shaped and reshaped the geologic and natural features of the Northeast; how the Earth's climate, animals, and plants are intertwined with and, in fact, controlled by the geologic evolution that shapes our planet; and how the Northeast fits, in time and space, into an everchanging picture of global evolution.

One author of this book first studied geology in the 1950s; the other came to the study of geology in the 1970s. During the two decades that separated our introductions to earth science a revolution occurred, as sweeping in its consequences as any other in the history of science. The father had no chance to pass down his geology texts to the daughter. By the time the daughter went off to college, the textbooks used by the father were obsolete.

Nowadays, if you pick up a geology text from thirty years ago, you have the sense that you are handling some quaint artifact from the distant past. Amid today's exciting and dynamic theories, the old texts seem brittle as fine antiques. The nature and the scope of the revolution in geology are described in the pages that follow. Let us say that the Earth as described in the father's texts and the Earth as we understand it today are as different as a frozen lake and a wild river. Thirty years ago, the Earth's crust was judged to be essentially static; today, geologists describe a planet with a mobile and dynamic crust, an animated being—moving, churning, rising, sinking, breaking, folding. In the 1950s you would have had difficulty finding a scientist who believed that the continents roam laterally across the face of the globe; today you would have trouble finding one who does not.

The continents move! They drift across the face of the Earth, sometimes colliding, sometimes rifting apart. The ocean floor is continually recycled through the interior of the Earth. A map of the Earth's surface today bears little resemblance to the surface as it was when dinosaurs first roamed the planet. The new geology of drifting continents and recycling oceanic crust provides fresh insights into the making of the Earth's landscapes.

The history describing the northeastern United States, the landscape the authors know and love best, is just a small part in a grander, global story about crustal transformations, one chapter in the story of a planet that is dynamically alive. We decided to tell the story of the Northeast as defined by the new geology because it has not been done before. Many excellent books have been written about the landscape of our region. A few that come to mind as classics in their field are Neil Jorgensen's *A Guide to New England's Landscape,* Arthur Strahler's *A Geologist's View of Cape*

Cod, George Bain and Howard Meyerhoff's *The Flow of Time in the Connecticut Valley*, and Christopher Schuberth's *The Geology of New York City and Environs*. Other useful books are listed in the bibliography. We have made use of all of them in telling our story. These other works, though, either have a local character or they were written before the revolution in geology. It is our intention to tell the story of the entire region—from New Jersey to Maine, from the Adirondacks to Cape Cod—and to tell it in the language of the new global geology. We present a full sweep of the region's geological history—from the time before the most ancient mountains were lofted skyward, to the final and fleeting signature that human beings have put upon the land. At every step in the story we endeavor to show how familiar landscapes came to be. Each chapter (except the two introductory chapters) begins with a global map that shows the positions of the continents at the time of the events described. Each chapter concludes with a brief summary. The map is an invitation to explore the Earth's dynamic past; the synopsis is a chance for reflection. We have tried to make the rhythm of the book reflect the rhythms of the Earth itself.

Throughout the book we draw on the most recent research and interpretations of the complex geological history describing the Northeast. Inevitably, some of the "facts" we present will become dated. New research and data will lead to better understanding of the rock record and to reformulation of prior beliefs. We hope our presentation conveys a sense of the excitement that pervades geology today.

The events we describe are grand events—collisions among continents, oceans disappearing and coming to be, towering mountain ranges raised and erased, ruptures in the Earth's crust that flood the land with lava, chains of volcanic islands crushed against continents. The events are grand, but the features of the landscape created by these events are as familiar as the hill behind your home, the rocks in the stone walls that enclose your yard, and the grains of sand on the beach where you go to swim. The history of the landscape is written in *all* the rocks and stones of the Northeast.

During the long time that the Northeast landscape was being shaped, life was evolving. Our story begins at a time when the only life on Earth was microscopic, living in the sea. We recount the rise and fall of the trilobites. We tell how plants and animals came to invade the land. We trace the rise to prominence of the dinosaurs, and tell about their perplexing disappearance. And, at the very end, we will consider our own relationship to the landscape we have inherited from the past.

We have set out to tell an intimate and accessible story that captures the excitement that is geology, a science envisioning the Earth's evolution as a process without end. Your response to the pages that follow will determine whether we have succeeded. Above all else, we want our story to evoke new affection for a varied and beautiful landscape that was long and vigorously in the making.

Prelude

At 4:11 in the morning on November 18, 1755, the citizens of Boston were awakened by a sound like distant thunder. Then came the shaking. Windows rattled. Beams creaked and cracked. The floors of homes lifted and fell like rafts on a rolling sea. People already out of bed found they had to hold on to something to avoid being thrown to the ground. Trees swayed as if in a hurricane. Weather vanes were broken from rooftops and chimney pots tumbled to the ground. Church bells pealed in their towers as if calling worshippers to service. In the harbor fish were killed by the thousands, and people on ships felt as if their vessels had struck bottom.

The earthquake epicenter, not far north of the city, was near Cape Ann. The shock, felt as far away as Chesapeake Bay and Nova Scotia, was the most severe earthquake to affect the northeastern United States in historic times. Somewhere deep in the Earth's crust an old fault had given way: Rock slipped against rock in a sudden, terrible shift, and all New England rang like a bell.

The geological circumstances that led to the great Cape Ann earthquake of 1755 had their origin long before the first European colonists came to these shores, and even before the arrival of the native Americans. Boston preachers thundered that the quake was divine retribution for a people's sins, but no amount of godly living could have prevented it. The stress that was relieved in one powerful shift of rock had been building under Cape Ann far longer than the few thousands of years that the preachers allotted to the age of the Earth. The earthquake, a shock from deep geologic time, echoed across hundreds of millions of years.

Chapter One

A Geological Revolution

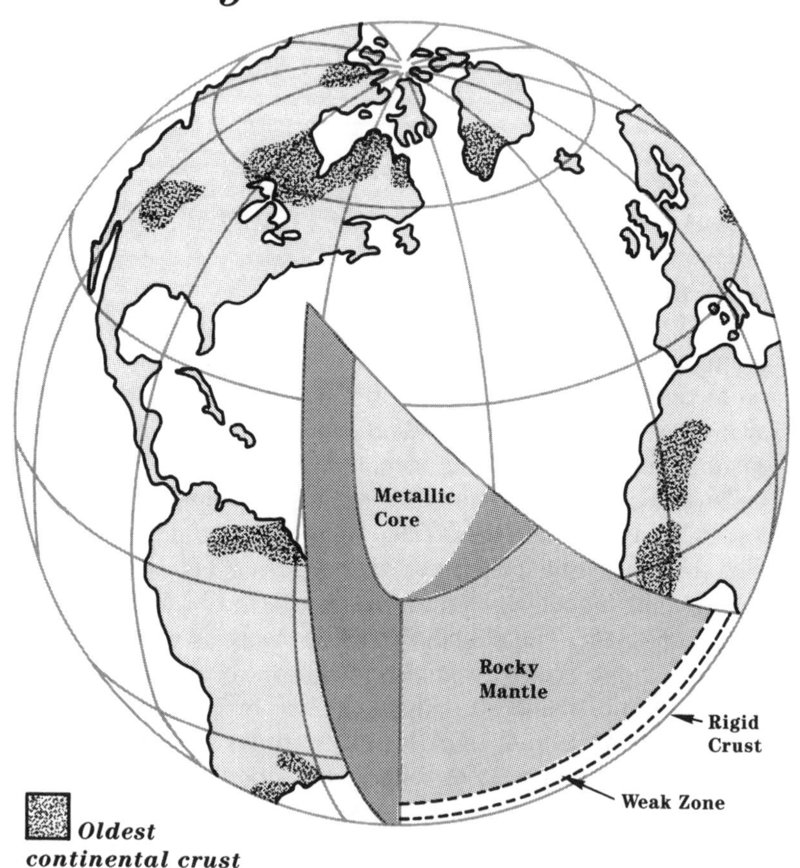

Precambrian	Paleozoic	Mesozoic	Cenozoic

4600 545 445 355 250 145 65 2

Millions of Years Before Present

Scale not proportional

In Boston today one can find rocks that formed from lavas poured out by ancient volcanos. The red sandstones in Connecticut were deposited in now-vanished deserts. New Hampshire receives its Granite State nickname from masses of rock that were once underground pools of molten minerals. Some mountains in eastern New York appear to have been pushed across the border from Vermont. The rocks in upper New York state enclose fossilized tropical coral reefs. Throughout the entire region exposed outcrops of rock bear signs that they were scoured and scraped by glaciers.

The rocks in the Northeast tell of dramatic events in the region's past. On at least three occasions, compressive forces within the earth squeezed up towering alpine ranges of snow-capped peaks that stretched from New Jersey to Maine and beyond. On another occasion, the region was stretched horizontally, opening gashlike rifts in the crust that oozed lava. At times in the past the climate in the Northeast was like Nevada's; at other times, like Antarctica's. Only twenty thousand years ago the entire Northeast was mantled by a layer of ice a half-mile thick. All this history and more is recorded in the rocks. In these first two chapters, we look at some of the geological ideas that enable us to read the story that is written in stone.

In general outline, the Northeast's geologic history has been known for a century. But even though we have known roughly *what* occurred and *when*, a satisfactory explanation of *why* these events occurred was lacking. The *why* is now available. The years since 1962 have witnessed a revolution in geology. A new and comprehensive theory has utterly changed our way of thinking about the Earth. For the first time, geologists have a global view of the forces that shape the Earth's crust.

The new theory is called *plate tectonics*. The word "tectonics" refers to the forces that *build* and *change* the surface of the Earth (it has the same root as the last half of "architect"). The meaning of the word "plate" will soon become clear.

Continental Drift

When the first reasonably accurate maps of the Atlantic Ocean were drawn late in the sixteenth and early in the seventeenth century, perceptive observers saw that the two sides of the ocean matched like the pieces in a jigsaw puzzle. With a little snipping and bending at Panama, South America and North America could be slid eastward and joined with Africa and Europe, with few gaps or overlaps. In our own century, the fit became more remarkable when ocean-floor maps defined the pieces of the puzzle by the continental slopes (where the continents precipitously fall away to the ocean abyss) rather than by the shorelines (where water meets land). When we assemble the jigsaw (Figure 1.1), the ocean simply disappears. The bulge of Brazil nestles trimly in the bight of Africa. The continental slope off the northeastern United States fits snugly against the coast of northwest Africa. On this map of joined continents, the trip from Maine to Morocco is a matter of a few hundred miles.

Is this jigsaw fit between the continents merely a coincidence, or were the eastern and western continents once joined? One of the first to ask this question scientifically was Alfred Wegener, a German meteorologist. In his *Origin of Continents and Oceans*, published in 1915, Wegener argued that the continents were once united in a supercontinental land mass he called Pangaea ("all-Earth"). Wegener supported his assertion with more than the jigsaw model uniting the continental margins. He showed that similar rock formations are found on the eastern coast of South America and the western coast of Africa. And he pointed out the similarities in the fossils found in rocks of South America and Africa. Not only did the *shape* of the puzzle pieces match, but also the *picture* on the puzzle.

Wegener believed that the continuity of coastlines, rocks, and fossil assemblages was more than a coincidence. He proposed that, at some time during the Mesozoic Era (which lasted from about 250 to 65 million years ago),

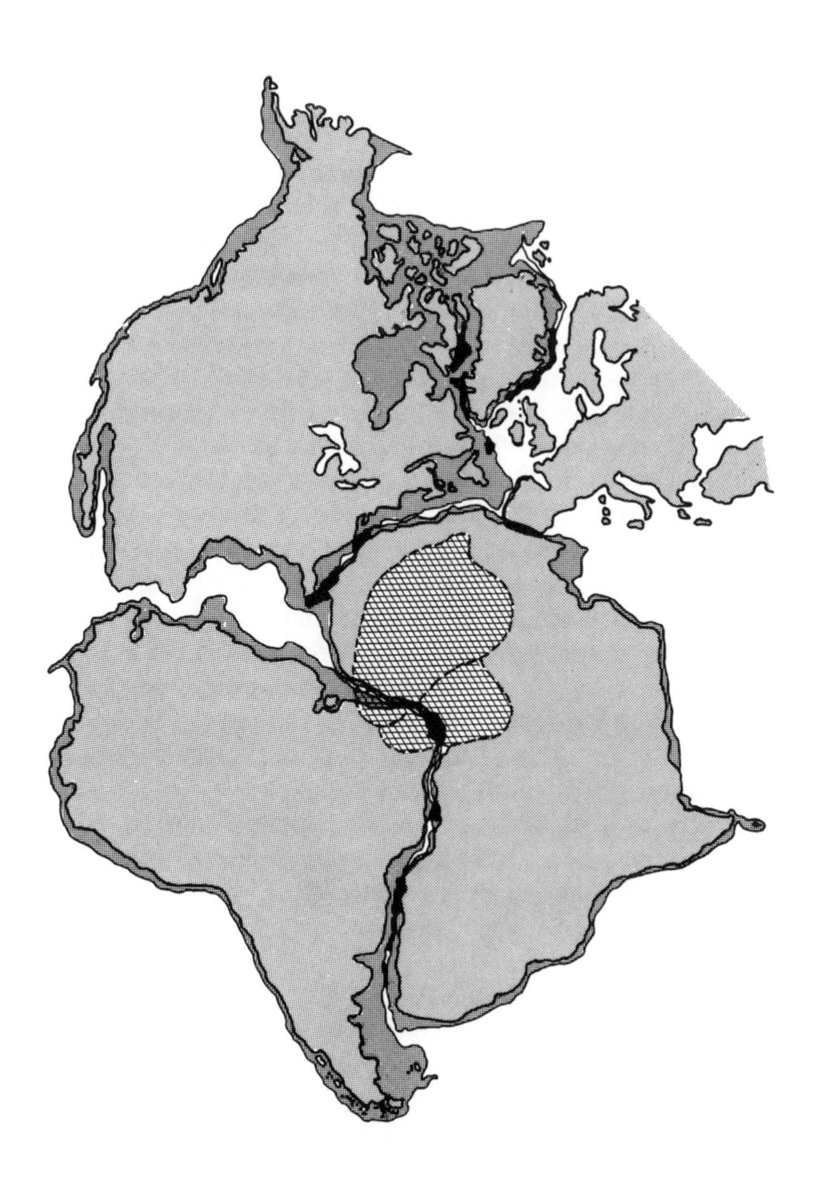

FIGURE 1.1—This "jigsaw puzzle" of the Atlantic continents is based on a reconstruction by Sir Edward Bullard, J. E. Everett, and A. G. Smith of Cambridge University. The cross-hatched areas in South America and Africa show matching rock formations more than 200 million years old.

Pangaea broke apart and the continents subsequently drifted to their present positions. He believed the light, granitic continents drifted like rafts across and through the denser rock that composed the ocean basins. The theory had just two problems: How could continents push their way through rigid rock? And what was the force that propelled them? Wegener thought that the continent-dispersing force had something to do with the Earth's rotation. But physicists objected that any force exerted by the Earth's rotation was much too weak to push continents through solid rock.

In the decade after Wegener's book was published, his theories were examined by the world's leading geologists and geophysicists—and almost unanimously rejected. The apparent rigidity of the Earth's crust and the absence of a propelling force seemed insurmountable problems. The idea of continental drift fell into disrepute. Wegener died in 1930, at age fifty, during a meteorological expedition to the Greenland ice cap. He would have been in his eighties when his theories about moving continents and a unified Pangaea were dramatically revived in the 1960s.

Wegener's ideas found renewed interest with developments in several areas of earth science. One of these was extensive exploration of the ocean floor; another was research in terrestrial magnetism.

Geopoetry on the Sea Floor

The 1950s were a "golden age" for sea-floor exploration. New technologies such as echo sounding, or sonar, made it possible to map the sea floor from surface vessels. Dredging and drilling techniques were devised for recovering rocks and sediments from the bottom. Among the many discoveries in the 1950s and early 1960s were two of particular interest. One had to do with the *topography*, or physical shape, of the sea floor, the other with its *age*.

A remarkable system of ridges was discovered on the floors of all the ocean basins. If the waters could be drained from the oceans, this world-girding mountain range would be the most imposing geographic feature on the face of the planet. In the Atlantic, the ridge lies exactly at the middle of the ocean basin and parallel to the margins of the continents. The crest of the ridge is *rifted,* as if sliced along the top by a knife. It is also warmed by an atypically large outflow of heat from the Earth's interior. Finally, the rocks on the ocean floor are youngest at the ridge (brand new!) and oldest near the continental margins. Nowhere does the ocean floor appear to be older than a few hundred million years.

Putting all these facts together, American geologist Harry Hess in 1962 came up with an idea he called *sea-floor spreading,* a variation on Wegener's continental drift. According to Hess, the oceanic crust is continuously extruded at the mid-ocean ridges, oozed up from Earth's hot interior. As new crust is created, the ocean floor is pushed away from the ridges, symmetrically to either side (Figure 1.2a). Slowly but inevitably, the ocean basins widen as the Earth's crust grows at the ridge.

And yet, the Earth's crust is not increasing in area. If crust is being created at the ridges, then it must be destroyed elsewhere. Hess suggested that crust is consumed at ocean trenches, those deep, curving canyons in the sea floor that are today found mainly in the western Pacific. At the trenches crust is forced back into the Earth for recycling (Figure 1.2b). Every few hundred million years *the entire oceanic crust of the Earth is cycled through the Earth's interior!* These great looping motions in the planet's material substance are driven by convection in Earth's interior—that is, by the tendency of hot matter to rise and cool matter to sink. Not only is the crust dynamic, the body of the Earth is in motion, slowly churning in vast convective loops.

Hess's sea-floor spreading hypothesis was so revolutionary that at first even he called the idea "geopoetry." But the hypothesis had two things going for it that Wegener's ideas

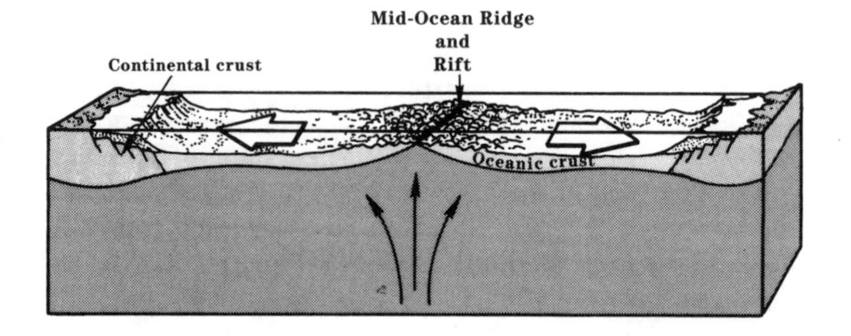

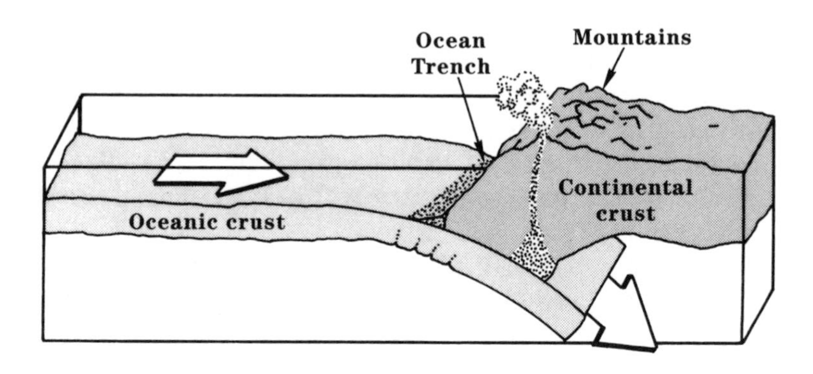

FIGURE 1.2 a—A cross-sectional view of a mid-ocean ridge shows hot mantle material being extruded along the axis of the rift valley. White arrows indicate direction in which oceanic crust moves away from the ridge. b—At the ocean trenches old oceanic crust is subducted back into the interior of the Earth by being forced under continents or other slabs of oceanic crust.

lacked. First, it provided a physical mechanism—convection—to account for lateral (sideways) motions of the Earth's crust. And second, the continents were not required to plow through solid rock. Rather, they moved *with* the moving crust, riding it like cargo on a raft. The rock of which the continents are made is too light to be dragged back into the denser interior (just as it is difficult to push a block of balsa wood under water). As the planet's denser oceanic crust is recycled, the lighter continents are dragged hither and yon, smashed together and rifted apart, sometimes growing as new light materials rise from the interior, but staying always on the surface of the planet.

Fossil Magnetism

As oceanographers were drawing new maps of the sea floor, geologists on land were making exciting discoveries about the Earth's magnetic field. It turned out that magnetism was the key to proving that continents drift.

Almost everyone at some time in their schooling sees iron filings sprinkled around a bar magnet. Every filing acts like a tiny compass needle, lining up with the magnetic field of the magnet. The sprinkled filings make that field visible, and we see lines of magnetic force "flow" from the south pole of the magnet to the north pole, in curved loops (Figure 1.3a). The Earth's magnetic field is similar to that of a bar magnet, as if a giant bar magnet were buried deep in the planet's core. But no geologist believes a solid bar magnet *really* lies there. The core is mostly iron, but it is too hot (in fact partly molten) to be magnetized in the ordinary way. A normal bar magnet loses its magnetism at high temperatures. Almost certainly, electric currents flowing in the molten part of the core generate the magnetic field. The Earth is an electromagnet.

A compass needle at the Earth's surface lines up with the Earth's magnetic field like the iron filings near a bar magnet. That is why a compass needle points north. The needle also

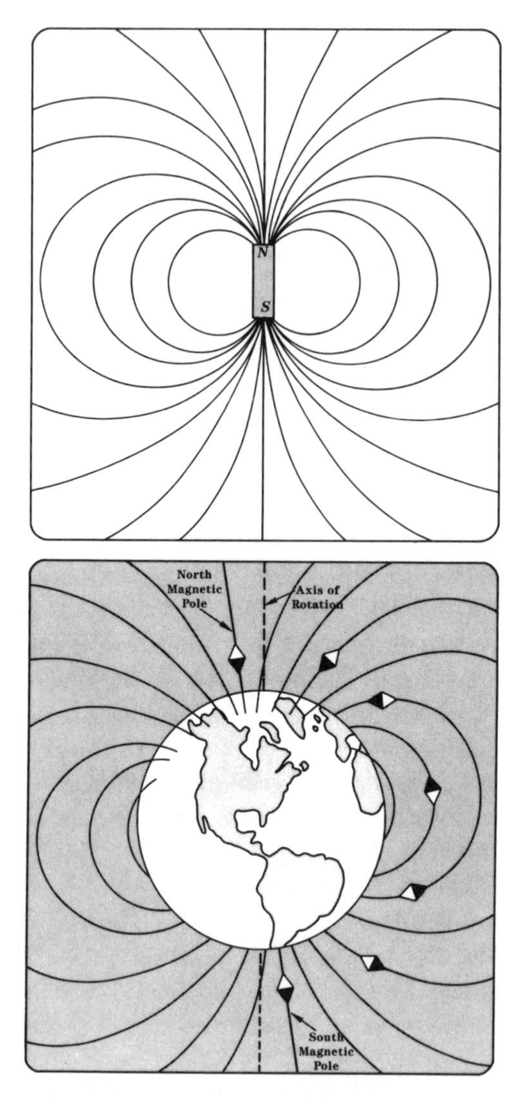

FIGURE 1.3 a—As this illustration suggests, the field of a bar magnet can be made visible by a sprinkling of iron filings. **b**—The magnetic field of the Earth is similar to that of a bar magnet. The magnetic poles of the Earth are near the poles of the Earth's rotation. The "compass needles" indicate the direction and inclination (dip) of the field observed at different points.

has a *dip,* or angle, into or out of the ground (Figure 1.3b). Near the north and south magnetic poles, the needle dips directly into or out of the ground; near the equator it lies parallel to the ground (no dip). The dip of the needle depends on latitude—how far you are north or south of the equator.

Under specific conditions, rocks can record and retain a trace of the orientation that the Earth's magnetic field had (both direction and dip) at the place and time the rocks formed. Lava includes iron-based minerals that at some temperatures will respond like iron filings to the Earth's magnetic field. While the solidifying lava is still hot (but not *too* hot), the iron-based mineral grains will line up with the Earth's field. When the lava hardens into solid rock, the current orientation of the mineral grains is frozen. This "fossil" magnetism in volcanic rocks, called paleomagnetism ("paleo" = "old"), records the magnetic field at the time and place where the rock formed. Marine sediments too can retain a paleomagnetic record of past fields. As tiny iron-based mineral grains eroded from the continents drift down through sea water they often turn so that they line up with the Earth's field. When the grains settle on the sea floor, as part of growing beds of sediments, they retain that orientation. In this way, sediments on the sea floor, or sedimentary rocks of marine origin, record information about past magnetic fields.

By the late 1950s, geophysicists had devised sophisticated instruments that enabled them to discover and read the weak paleomagnetic record in the rocks. By studying the magnetic direction and dip in continental rocks, they were able to determine where the Earth's magnetic poles were in relation to the rocks at the time the rocks formed. They were surprised to discover that the north and south magnetic poles had apparently drifted widely across the face of the planet. And more disturbing, ancient rocks *of the same age* on opposite sides of the Atlantic Ocean indicated different positions for the poles.

North American rocks that are 300 million years old

have zero dip, suggesting that 300 million years ago North America lay near the *magnetic* equator. Those rocks point to a north magnetic pole somewhere in the present Pacific Ocean south of Japan (Figure 1.4a). Geophysicists believe, however, that the electric currents in the core that maintain the Earth's magnetic field are driven by the planet's rotation; if they are, then the magnetic poles should never drift far from the geographic poles (which define the axis of the Earth's rotation). But if the magnetic poles didn't drift, what did? The answer: the continents, just as Wegener had guessed! Three hundred million years ago the north magnetic pole was not in the Pacific Ocean south of Japan; it was close to where it is today. It was North America that had a different location and orientation. North America *was on the equator and tipped on its side* (Figure 1.4b).

Another magnetic paradox was resolved by the theory of continents drifting—rocks on opposite sides of the Atlantic that are more than 200 million years old point to different poles. But when we reassemble the eastern and western continents into Wegener's jigsaw supercontinent, Pangaea, the paleomagnetic poles, as deduced from ancient rocks on the different continents, match up perfectly. Conclusion: *The continents were not always where we find them today.* Since 1958, the thousands of paleomagnetic measurements collected from rocks of all ages, all around the world, have allowed geologists to construct maps of the continents at various times in the past. These maps, used in the chapters that follow, vividly depict how the Earth's surface has continually evolved.

Plate Tectonics

For Hess's idea of sea-floor spreading to work, the rock in the Earth's interior must be able to *flow* in convective loops. Analysis of earthquake waves reveals a picture of the Earth's interior that is consistent with Hess's hypothesis. The rigid, brittle crust of the Earth is as thin compared to the whole planet as the shell to the whole egg. Only this

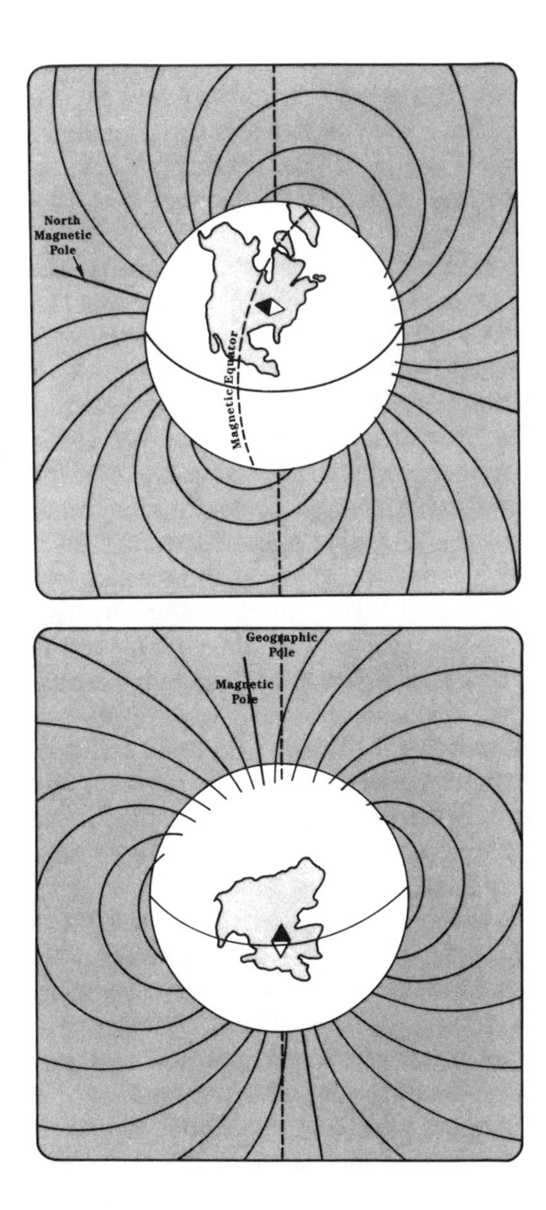

FIGURE 1.4 **a**—The magnetism of 400-million-year-old rocks in North America suggests that the positions of the Earth's magnetic poles have shifted dramatically in the past. **b**—Actually, however, it is the continents that have changed positions, not the magnetic poles.

outer shell is rigid enough to fracture and slip violently; no earthquakes have been recorded in the planet's interior. The inside of the Earth is so hot (from heat generated by the decay of natural radioactive minerals) that the rocks are ''soft'' and plastic.

The cool, rigid ''eggshell'' of the Earth is called the *lithosphere*, from the Greek word for ''rock.'' The lithosphere is thickest under the continents—and especially thick where mountains stand—and thinnest under the ocean basins. Under the lithosphere the body of the Earth is too hot to be brittle. The interior rock is solid, but hot enough to flow under the influence of gravity, just as the iron from a blacksmith's forge is hot enough to flow under the force of the smith's hammer. The layer of the Earth just under the crust is particularly weak and close to the melting point, and in some places molten. This layer in the Earth's interior is called the *asthenosphere*, from the Greek word for ''weak.''

The plastic rock below the lithosphere churns in convective motions, driven by heat and gravity like pudding in a pan on the stove. ''Churning'' is a vivid and accurate word for the asthenosphere's convective motions, but perhaps misleading. The motions are slow on the human scale—a few centimeters per year. These churning motions dragging on the Earth's eggshell crust break it into thin rigid sheets, the *plates*, which are propelled about the Earth's surface (Figure 1.5). Where the plates are cracked and pulled apart, new crust is added by volcanic action. This activity occurs at the ocean ridges. At other places, plates are pushed together. When plates composed of ocean crust converge, one plate must be forced under the other, back into Earth's interior. This sinking back into the Earth, called *subduction*, creates deep ocean trenches. A rigid subducting plate will bend and break, generating earthquakes. The pressure, friction, and chemical transformations to which rocks are subjected during subduction generate heat, and a line of active volcanos invariably stands above a subducting plate (as along the western margin of the Pacific Ocean).

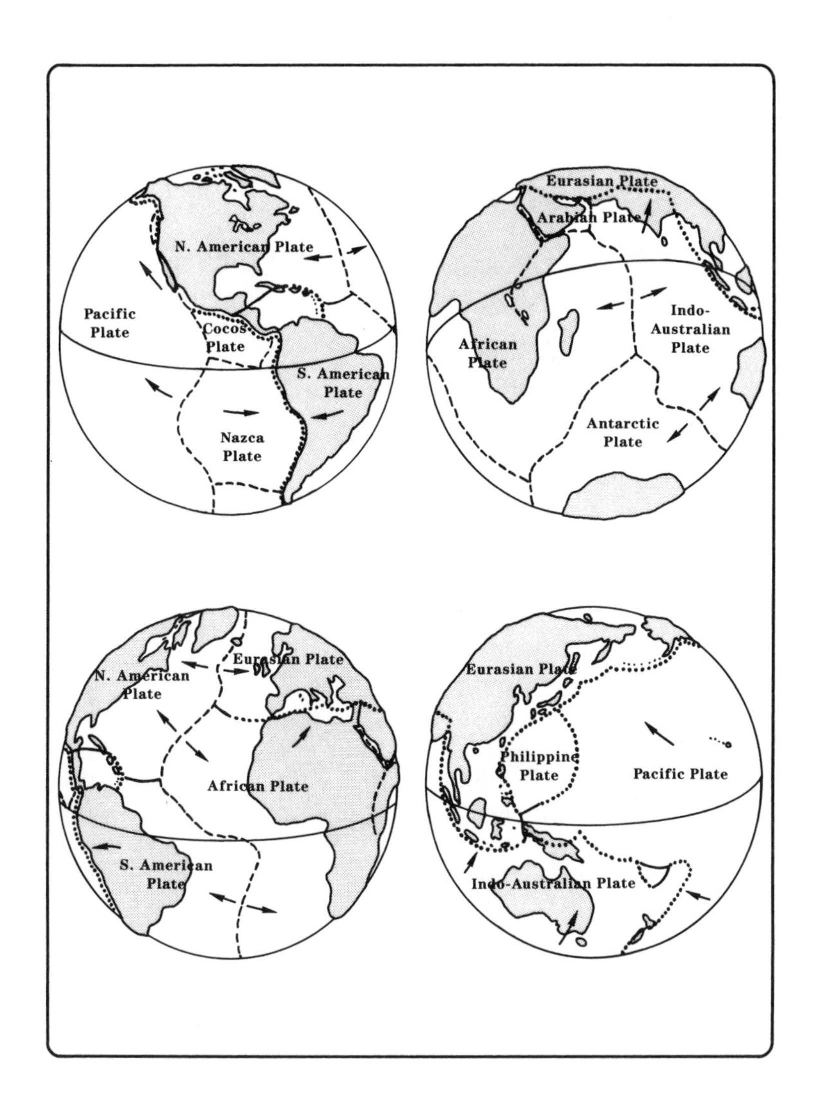

FIGURE 1.5—Four views of the Earth show the pieces of broken "egg-shell" called *plates*. The dashed lines are the mid-ocean ridges where new crust is being created. The dotted lines are places of plate convergence where old crust is being subducted.

The continents ride the moving plates. When oceanic crust and continental crust collide (as they are presently doing along the western coast of South America), it is always the denser oceanic crust that sinks back into the Earth; the continental crust is crumpled and mountains are raised (Figure 1.2b). When continental crust meets continental crust (as where India is presently colliding with Asia along the line of the Himalayas), the crust has nowhere to go but up. Almost all the earthquake and volcanic activity on the face of the planet occurs along plate boundaries. The plate boundaries are the ''cracks in the eggshell'' where new crust is made or old crust consumed.

A Dynamic Earth

The crust of the Earth is dynamic. The continents drift, arranging and rearranging their bulk. The floors of the oceans are recycled. The entire planet is always moving—spreading, sinking, breaking, folding. This is the central tenet of the new geology: We live on the thin, fragile skin of a planet that is—metaphorically speaking—alive!

Hess's idea of sea-floor spreading quickly passed from ''geopoetry'' to ''geofact.'' Sea-floor spreading was soon transformed into the comprehensive theory of plate tectonics. Confirmation for the theory suddenly seemed to appear everywhere. By the end of the 1960s the revolution in geology was essentially complete. We cannot here rehearse the myriad ways in which the new theory has been confirmed, or the ways in which it continues to illuminate our understanding of the planet's face. Our primary task is to understand how the landscape in the Northeast came to be. We will trace the slip and slide of plates, the riftings and collisions of continents, which gave our landscape its form. The story that unfolds in these chapters illustrates the enormous power that plate tectonics has in making sense of the geological past.

Chapter Two

The Writing in the Stone

Millions of Years Before Present	Eras	Periods	Significant Events
2	Cenozoic	*Quaternary*	At least four ice ages occur Appearance of humans
65		*Tertiary*	Beginning of the Andes Grazing animals multiply Grasslands spread Collision Africa and Europe, beginning of the Alps Collision of India and Asia, beginning of Himalayas Mammals and birds flourish Rocky Mountain uplift begins
145 205 250	Mesozoic	*Cretaceous* *Jurassic* *Triassic*	Extinction of dinosaurs Flowering plants First mammals Age of the dinosaurs First birds Opening of present Atlantic Ocean Pangaea splits up Cone-bearing trees Reptiles established
545	Paleozoic	*Permian* *Carboniferous* *Devonian* *Silurian* *Ordovician* *Cambrian*	Amphibians established Collision of Europe and Asia, beginning of Urals Collision of Africa and North America, beginning of Appalachians Extensive coal-forming forests Winged insects Plants and animals move onto land First vertebrates, the fishes, appear First animals with shells
4600	*Precambrian*		Sexual reproduction Respiration Photosynthesis Growth of continents Origin of life Crust forms; first oceans and atmosphere Differentiation of core and mantle Earth forms from solar nebula
	Scale not proportional		

Geologic maps are objects of great beauty—multihued, color-splashed, polychromatic. The colors on a geologic map indicate rocks of different types and ages as they are exposed at the surface of the Earth. Imagine that you could scrape bare the surface of the Earth, down to solid rock, down to bedrock. Scrape away the works of humankind—the cities, the roads, the orchards, the stone walls. Scrape off all plant life—the prairies, the bogs, the forests. Scrape off the unconsolidated soil, sand, gravel, and erosional debris. Scrape down to the bare, stony skin of the planet. Now, examine the rock at your feet. What kind of rock is it? How old is it? Sort all of this out as best as you can and then *map* the crust. Use color. For rocks that once existed in the molten state, use the reds and oranges we associate with fire. For rocks that have been transformed by heat and pressure, use tints of brown and gray. For rocks formed from consolidated sediments—sandstones, mudstones, limestones—color your map with violets, blues, greens, and yellows, according to the age of the sediments. When you have finished, you will have a map as gaudily and gaily colored as a painting by a madcap expressionist.

Geologic maps representing some parts of the Earth's surface are rather plain things. A geologic map of Iowa requires only a few colors—the colors of sedimentary formations—arranged in more or less parallel bands, like a wavy flag. A geologic map of the northeastern United States is rather different. Here the colors are more varied and jumbled than in any other part of the continent. At first glance the map gives the impression of an artist's palette on which *all* the colors have been swirled and mixed together—streaks of green and brown, loops of pink and orange, striations of blue and purple, great dollops of red—helter-skelter, slapdash, chaotic. Sorting it all out, reading a consistent story in this twisted pastiche of color, can seem an impossible task.

Yet a closer look at a geologic map of the Northeast reveals something less than total chaos. Patterns and trends become apparent. Adjacent colors appear in similar sequences on different parts of the map. The swirls of color have a linear trend—northeast to southwest—roughly parallel to the coast. The colors that represent volcanic rocks are confined to two regions, in northern New Jersey and central Connecticut. The granites' scarlet reds are clustered in New Hampshire and Maine. But, before we can interpret the patterns on the map, we need to understand how different rock types are formed and how geologists determine their ages.

Rocks

From among the rocks in the northeastern United States you could easily compile a satisfactory collection of all the major rock types in the Earth's crust: old rocks and young rocks, hard rocks and soft rocks, rocks formed in the fiery furnace of the Earth and rocks consolidated from the scrapings of glaciers, rocks consisting of compacted volcanic ash and rocks laid down on the floors of seas, rocks enfolding skeletal remains of microscopic marine organisms, and rocks impressed with tracks left by dinosaurs. All these can be found in the Northeast.

Since the earliest days of their science, geologists have recognized three basic types of rock—igneous, sedimentary, and metamorphic—which differ in their mode of origin.

Igneous rocks: Igneous rocks once existed in a molten state at very high temperatures. They form as molten minerals cool and harden, exactly as water freezes to become ice. In these rocks, grains of pure minerals—such as quartz, feldspar, or mica—interlock like the pieces that form a parquet floor. The size of the grains indicates how quickly the rock cooled from the liquid state; smaller grains suggest faster rates of cooling.

Intrusive igneous rocks cool and solidify slowly within the body of the Earth. Molten rock in Earth's interior is called *magma*. Granite, the most familiar intrusive igneous rock, appears in vast outcroppings throughout the Northeast, chiefly in New Hampshire and central and southern Maine. From these rocks New Hampshire gets its nickname, "the Granite State." The mineral grains of granite, easily visible, give the rock its characteristic speckled appearance. Because of its large grain size, we know that granite typically forms from pockets of magma several miles below the Earth's surface, where heat is not readily dissipated. Yet we now find these rocks exposed on the surface. This exposure implies that the crust has been lifted, and the overlying rocks that originally buried the granites have been stripped away by weathering and erosion. Formation, lifting, and dissolution of the granites are important chapters in the geological history of the Northeast.

The second kind of igneous rocks are *extrusive*. The rapid cooling that made these rocks could only happen on the surface of the Earth, where heat is quickly dissipated. Molten rock on the surface is called *lava*. The mineral grains in extrusive rocks are generally too tiny to be distinguished with the naked eye. Volcanic glasses such as obsidian cooled so fast that *no* mineral grains had time to grow. Wherever you find extrusive igneous rock you are looking at evidence of volcanic activity. Throughout the Northeast, extrusive volcanic rocks attest to a fiery, eruptive history. In Brookline, near Boston, you can find rocky "bombs" blasted from volcanoes with shattering violence. In central Massachusetts, near Amherst and Granby, you can walk on fields of compacted ash and lava. The Watchung Mountains in New Jersey are carved from basalt, hardened lava that poured through rents in the crust and flooded the land with an incandescent sea. An account of these spectacular volcanic events is part of our story.

Sedimentary rocks: Sedimentary rocks tell a gentler tale. Although igneous rock typically is created along with the tectonic activity that builds mountains and pulls continents apart, sedimentary rock is often associated with the long weathering and erosion that tear mountains down. The sand grains on Long Island and Cape Cod beaches are made up of tiny crystals of quartz that were once part of the inland New England granite rock. Wind, rain, freezing and thawing, and the grinding force of glaciers broke the granite apart. Some of the minerals in the granite, such as feldspar, readily dissolved in water and became mud. The more resistant quartz grains remained intact. Flowing water carried the decomposed granite downhill to low-lying basins, where it was deposited as mud or sand. Today's mud flats and glistening sandy beaches along our northeastern shore are the detritus or leavings of mountains. Eventually, over millions of years, these sands and muds will accumulate in great thicknesses on the sea-covered continental margin, and pressure and chemical cementation will turn them into sandstones and mudstones.

Plate-tectonic movements of the crust can lift sedimentary rocks from the low-lying basins where they form to the tops of high mountains. The Catskill Mountains in New York are composed of sedimentary rock. On the shore of the Ashokan Reservoir in the Catskills you can pick up chunks of sandstone enclosing the fossil shells of creatures that lived long ago on an ancient sea floor. But where sedimentary rocks are lifted into mountains, they are quickly cut down. As a rule, sedimentary rocks erode more easily than igneous rocks, and for this reason they are often found to underlie valleys; the highest peaks in the Northeast—the White Mountains, Green Mountains, and Katahdin in Maine—are composed of resistant igneous and metamorphic rocks. Exceptions to this rule are the bluffs in the Shawangunks (pronounced "Shon-gums"), at the southern end of the Catskills. They have retained their imposing identity because they are composed of a particularly resistant sedimentary rock, the so-called Shawangunk conglomerate, consisting of solidly cemented quartz pebbles.

Sediments, such as sand or mud, accumulate on sea floors or inland basins as flat layers, or *strata*, and sedimentary rock characteristically retains the layering, called *stratification*. These rocks usually break along the strata, forming flat slabs. The Ice Caves, at the crest of the Shawangunks near Ellenville, New York, are made from great jumbled slabs of Shawangunk conglomerate that have fallen from the cliffs above. The strata in the Shawangunks lie flat; they were lifted almost straight up, without tilting, like the floor of an elevator. If the strata in sedimentary rocks are not horizontal, then we know that earth movements have tipped or folded them from their original orientation. Along the New York Thruway between Exits 21 and 21B one can see stratified limestone beds that have been wrinkled up and down like the folds in a rumpled carpet (Figure 2.1). The upfolds (arches) are called anticlines; the downfolds (sags) are called synclines. These folds are displayed most dramatically on New York 23A just west of the Thruway intersection. Not far away are mudstone beds that have been tipped almost vertically.

Unless beds of sedimentary rock have been entirely overturned, the lowest-lying strata are the oldest. A drive along New York 28 from the Hudson River near Kingston to Big Indian, a trip of about thirty miles, takes you up through successively higher, and hence younger beds of shale, limestone, siltstone, sandstone, and conglomerate, on a journey that traverses 100 million years of geologic time, from the time when trilobites dominated the seas, to the time when the first amphibians colonized dry land.

Sedimentary rocks are not limited to the detritus of erosion. Anything that can accumulate in a basin can become a sedimentary rock. Limestone consists of consolidated carbonate muds precipitated from sea water or of carbonate skeletal remains of marine organisms. Coal is made from deposits of organic material, the decaying plant matter found on the floors of swamps or bogs. Each kind of sedi-

FIGURE 2.1—An outcrop of folded limestone is exposed near Kingston, New York.

mentary rock tells a story about the environmental conditions at the time of its deposition. In the chapters that follow, we will look at the sands, muds, limes, pebbles, boulders, and coals in the sedimentary rocks of the Northeast and read there a history of ancient beaches, estuaries, seas, streams, mountain valleys, bogs, and swamps.

Metamorphic rocks: Metamorphic rocks are formed by *any* kind of rock (igneous, sedimentary, or metamorphic) altered by high temperature and pressure without remelting. Metamorphism does not change the rock's bulk chemical composition, but does produce new minerals and textures. Slate is a metamorphosed variety of shale, a sedimentary rock formed from clay; slate is common on the western side of the Taconic Mountains in New York and in north-central Maine. The famous Vermont marble is metamorphosed limestone. The gneiss and schist that underlie much of Manhattan are altered forms of sandstone and mudstone.

Frequently, metamorphism produces a thin, wavy *foliation* in rock; when you see metamorphic rocks exposed in a road cut, they will often look like the pages in a book seen edge on. Beautiful examples of foliation can be seen in road cuts near Burlington, Vermont, and along Interstate 91. Metamorphic rocks, wherever they occur, are evidence that the Earth's crust was squeezed and crushed. The Green Mountains and the Taconics consist primarily of metamorphic rocks. Each of these mountain ranges was created when the Northeast was locked in a vise grip and squeezed by colliding crustal plates. The rock that was originally at the core of each of these mountain ranges was transformed by heat and pressure into the metamorphic rock now exposed, millions of years later, at the Earth's surface.

Time

Kaaterskill Falls (Figure 2.2), just off New York 23A ("the Rip Van Winkle Trail") between Catskill and Saugerties,

FIGURE 2.2—Kaaterskill Falls, near Catskill, New York

New York, is the highest cascade in the northeastern region, higher even than Niagara Falls. It tumbles 260 feet across two great rocky steps. The "treads" of the steps are resistant sandstone. The "risers" are less resistant shale that has been eroded back under the falls, so that where the water leaves each step it flows over a protruding lip of sandstone. In its fall across 260 feet of sedimentary strata the water traverses millions of years in geologic time.

The sands and muds exposed at Kaaterskill Falls as sedimentary rock were deposited grain by grain in a low-lying basin by moving water, as even today the Mississippi River is building up a great sand and mud delta at its mouth. Grain by grain! How long did it take for beds of sand and mud 260 feet thick to accumulate? The strata exposed at Kaaterskill Falls are only a small section in a sequence of sedimentary formations, *thousands* of feet thick, which reach from the low-lying Hudson River Valley up into the Allegheny Plateau. Grain by grain these sands and muds and pebbles were eroded from mountains to the east, carried by water down the flanks of the mountains, and deposited in horizontal layers in lowland basins or shallow seas. All this deposition happened between 350 and 400 million years ago, at about the time the earliest land plants and animals were moving out of the sea. Deeper and deeper the sediments were buried as new sediments washed in from the highlands and the crust of the Earth sagged beneath their weight. Much later, after the sediments had been turned to stone, pressures in the crust lifted the entire region above sea level, and in some places squeezed the layers into great looping folds—synclines and anticlines. But the Catskill sandstones and mudstones and conglomerates were not lifted overnight. They rose millimeter by millimeter over millions of years. Their deposition and later rise cannot be measured on a human time scale. The events that ultimately formed Kaaterskill Falls required hundreds of millions of years for their completion.

James Hutton, a late-eighteenth-century Scottish gentleman farmer, taught us how to look at the layered rocks in the Earth's crust and see more than the several thousand years allotted to the Earth by Biblical history. Hutton's *Theory of the Earth*, read before the Royal Society of Edinburgh in 1785, conferred upon the planet sufficient time for mountains to be eroded away—grain by grain—and for mountains to be lifted—millimeter by millimeter—*by the same processes that we see occurring on the Earth today*. How much time? Hutton was not certain, although he knew it must be millions of years. His book was subtitled *An Investigation of the Laws Observable in the Composition, Dissolution, and Restoration of Land Upon the Globe*. What that complicated mouthful means is this: Look at the face of Kaaterskill Falls—the rock is apparently composed of cemented grains of sand and mud; the sand and mud must have been derived, as it is today, by erosion of mountains; the sand and mud must have been carried to lowland basins by moving water and deposited in horizontal layers; these rocks are today found at high elevations, meaning that the mountains have been restored by uplift. The time required for these events to unfold must have been incredible. And the process does not stop here. Visit Kaaterskill Falls today and you can see the water vigorously dismantling the Catskill Mountains. Slabs of rock that have fallen from the cliff face litter the foot of the falls; the slabs are being disassembled. Detritus from the mountain's face is carried by Kaaterskill Creek into the Hudson River, thence past New York City skyscrapers and beneath the Verrazano Bridge to the continental shelf, where it is deposited in beds growing in thickness that will first become sandstone and mudstone and then, in some future geologic era when the shelf is lifted and folded by new pressures in the crust, new mountains. In all this cycling of the substance of the Earth's crust, James Hutton claimed to see "no vestige of a beginning, no prospect of an end."

Following Hutton's lead, nineteenth-century geologists began to establish a time scale for the rocks in the Earth's crust. They could not tell the absolute age of the rocks, but they could determine relative ages.

First they invoked the principle of superposition, originally propounded by Danish physician Nicolaus Steno in 1669, which says that the lowest-lying strata in a sedimentary formation must have been deposited *before* the overlying strata. We assume that the rocks now exposed at the bottom of Kaaterskill Falls are older than those at the top. Of course it is possible that in places the tectonic upheaval of the Earth's crust was so great that a sedimentary sequence was *entirely overturned*, so that the oldest layers of rocks now lie on top. Such inversions are rare, and where they occur we can usually detect them by tracing the disposition of strata across broad geographic regions or by using clues preserved in the strata themselves, such as fossils, pits made by raindrops, ripple marks impressed by flowing water, or animal tracks and burrows.

Fossils are also effective clues for determining the relative ages of rocks. The story of rocks and the story of life's evolution are intimately related. Over the past billion years life has evolved from simple one-celled organisms to the wonderfully varied complex forms we know today. Along the way, huge numbers of species have flourished and become extinct. If similar fossils are found in rocks at different localities, then we can assume that those rocks are of the same age. Fossils of the same trilobite species—an extinct many-legged arthropod with a three-lobed shell—can be found at Hoppin Hill, North Attleboro, Massachusetts, and at the Fruitville Quarry near Lancaster, Pennsylvania; we can conclude that the rocks in the two localities are approximately coeval (both, in fact, are about 500 million years old). Geologists, paleontologists, and biologists work hand in hand to sort out the interwoven histories of life and rock.

As nineteenth-century geologists labored to construct a relative time scale for the world's rocks, they divided Earth's history into eras and periods, designations still commonly used today. The principal divisions in the time scale (for the past 600 million years) are the Paleozoic, Mesozoic, and

Cenozoic eras, for "old life," "middle life," and "recent life." The periods within these major divisions are often named for the place where the rocks of that age were first studied in detail. The Cambrian period takes its name from Cambria, the Roman name for Wales, and the Jurassic period from the Jura Mountains in France. The Cretaceous period derives its name from the Latin word for "chalk," a soft carbonate rock that is common among the rocks of that age. Very few sedimentary rocks on the face of the Earth are more than 600 million years old; the older sedimentary rocks have been removed by erosion or metamorphosed into new forms—and those rocks, where they are found, hold no fossils, at least none that are visible to the naked eye. That shadowy part in the Earth's early history was labeled simply Precambrian by nineteenth-century geologists. Our story of the Northeast begins in the late Precambrian, about 800 million years ago. This is by no means the true date at which the geological history of the Earth's crust began: the Earth is about 4.6 billion years old. But of the more than 3 billion years that separate the oldest rocks in the Northeast from the creation of the planet itself, we have only the murkiest glimpses.

The breakthrough in creating a *quantitative* timetable of geologic history came at century's end when radioactivity was discovered. Radioactive atoms are unstable and in time break apart into smaller entities. The precisely measurable rates at which radioactive elements decay can be used as a natural clock to determine the time gone by since rocks that contain those elements were formed. Radiometric dating makes it possible to put absolute dates on the geologic time scale. Over the years, many thousands of measurements have helped refine the age estimates shown on our time line introducing this chapter.

James Hutton's vision of deep time made modern geology possible. Earth's story requires for its appreciation that we stretch our ordinary human feeling for time. A million years is but a minute on the geologic clock. Tens of millions of years are required to create a range of mountains and wear them down. To tell how the Northeast landscapes were created, we first have to prepare our imaginations to range over eons.

One fine day in 1788, James Hutton took his friends James Hall and John Playfair on a boat ride along the Berwickshire coast in Scotland. They came to a place along the cliffs where almost horizontal strata of ''Old Red Sandstone'' were lying across the eroded vertical stumps of strata of an even more ancient ''schistus'' (Figure 2.3). Hutton outlined for his friends the long episodes of uplift and erosion, occurring in cycle after cycle, that were required to create this outcrop of stone. Grain by grain, sand was deposited and turned into stone; millimeter by millimeter, the sandstone strata were lifted and tilted and squeezed and turned into schistus; grain by grain the strata of schistus were truncated by erosion, and millimeter by millimeter the surface of the truncated layers subsided; grain by grain new strata of sand were deposited and turned into stone; millimeter by millimeter the crust rose and the sandstone was eroded. These long heavings and falterings of the Earth's crust culminated in the formation they saw before them. The Earth was ancient, Hutton asserted; how ancient he could not tell. John Playfair recorded his impressions from that day on the Berwickshire coast: ''On us who saw these phenomena for the first time, the impression will not easily be forgotten. . . . We felt ourselves necessarily carried back to the time when the schistus on which we stood was yet at the bottom of the sea, and when the sandstone before us was only beginning to be deposited, in the shape of sand and mud, from the waters of a superincumbent ocean. . . . The mind seemed to grow giddy by looking so far into the abyss of time.''

In the chapters that follow we will plunge into the abyss. Prepare to be giddy.

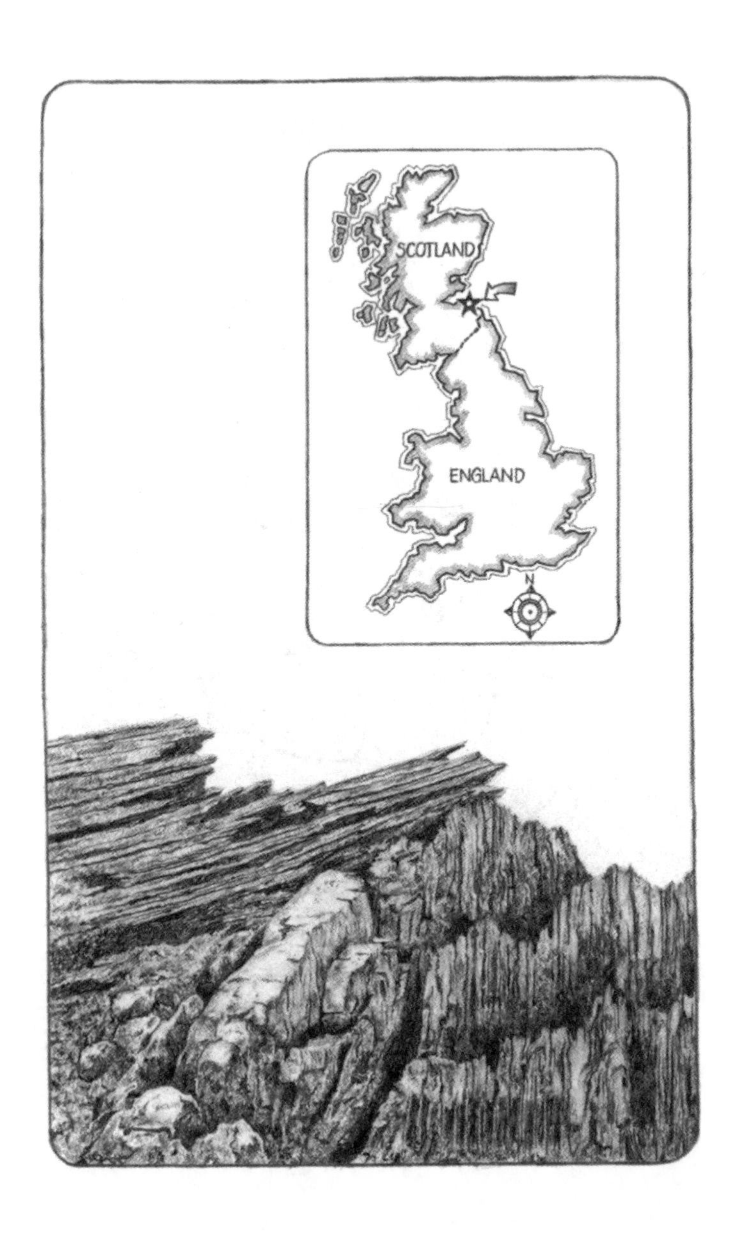

FIGURE 2.3—This outcrop at Siccar Point in Scotland depicts younger strata lying atop tilted and eroded beds of older rock.

Chapter Three

Laying the Foundation

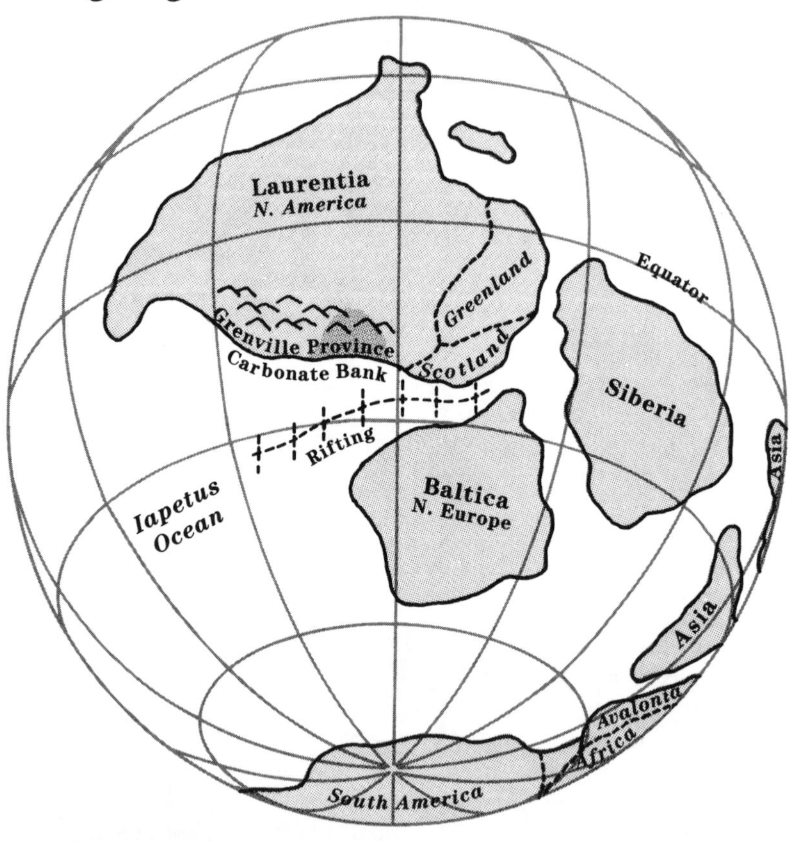

Precambrian	Paleozoic			Mesozoic		Cenozoic
4600 – 545	445	355	250	145	65	2

Millions of Years Before Present

Scale not proportional

"As old as the hills" is an expression often used to describe something of indeterminate antiquity. Can anyone doubt that by a human time clock hills are old? On a geologic time scale, however, the hills can be very young. Consider the hills around Boston Harbor that were the scene for early actions in the Revolutionary War. On Bunker Hill and Breeds Hill, just north of Boston, the colonists repulsed the British in the first important battle of the war, on June 17, 1775. George Washington made good use of the commanding ground of Dorchester Heights, a hill south of Boston, to emplace the battery of cannon with which he persuaded the British to evacuate the city. Breeds Hill, Bunker Hill, and Dorchester Heights are glacial *drumlins*, hills formed from gravel that was deposited under ice when New England was glaciated tens of thousands of years ago. Ten thousand years is just a tick of the geologic clock. The hills around Boston Harbor are geologically young, no older than the earliest archeological evidence for a human presence in the Northeast. About the Boston drumlins, a geologist might appropriately exclaim: "As young as the hills!"

Or consider the "Old Man of the Mountains," the famous granite face that looks down on Profile Lake in New Hampshire's Franconia Notch (Figure 3.1). Is the Old Man of the Mountains *old* by geological standards? Well, yes and no. The granite face in Franconia Notch has been recently shaped by erosion from ledges of Conway granite, yet this granite was emplanted under the ancestral White Mountains in Mesozoic times, about 200 million years ago. Two hundred million years is a considerable stretch of time even by geologic standards. The rock forming the Great Stone Face is therefore old; it had its origin deep inside the Earth at the time of the dinosaurs. Two hundred million years of erosion have cut away the ancestral White Mountains that once stood above the granite. As the overlying formations eroded the crust rebounded, just as a raft rises in the water when

FIGURE 3.1—The Old Man of the Mountains stands sentry duty at Franconia Notch, New Hampshire.

weight is removed. Eventually the deeply buried granite was exposed, and weathering began to carve the accidental features of the Old Man's face. The brow, the nose, and the chin are certainly no more than a few hundred years old. They are a brief, almost instantaneous snapshot in the mountain's geological history, precariously posed to go the way of the rest of the mountain. Without bolts and chains to secure them, chunks of the Old Man's face might already have fallen into the valley.

Just how old *are* the hills? The question is ambiguous. The answer depends upon whether we are talking about the hills as *topographic* features—areas of greater than average elevation—or about the rocks of which the hills are made. Topographically speaking, any hill is young. In a few tens of millions of years erosion can cut down even the highest mountain. Wherever a mountain range stands on Earth's surface, you can be sure that the crust has been uplifted in geologically recent times. But the age of the rocks themselves is another matter. Geologists date rocks from the time at which they formed—for igneous rocks, from the time they solidified from the molten state; for sedimentary rocks, from the time of their deposition; and for metamorphic rocks, from the time of their recrystallization. Like the Old Man of the Mountains, a topographic feature is usually much younger than the substance from which it is made.

Some hills that are both young and ancient are the Adirondack Mountains in New York. Topographically, the Adirondacks are among the youngest features in the Northeast. These wild and rugged mountains have been lifted and carved very recently on the geologic time scale. Indeed, they appear to be still rising! Yet the *rocks* in the young Adirondack peaks are among the oldest in the Northeast. They consist of hard, tough, metamorphosed granites that have been severely deformed by a billion years of tectonic activity.

A Window on the Past

The Adirondacks are something of a mystery. They are a young, growing mountain system in a part of the continent that is far from any active plate boundary, and should therefore be relatively stable. Geologists are not sure *why* the Adirondacks are where they are. Nevertheless, these splendid mountains provide us with a fortunate glimpse into the region's deep geologic past.

The Adirondack rocks date from the Precambrian era. They are close to a billion years old. Along with a few minor exposures scattered through the Northeast, they are the oldest rocks observable in the Northeast. Since the Precambrian, as the continents barged about the surface of the globe, sometimes colliding, sometimes being wrenched apart, these rocks have absorbed repeated shocks. Of the Precambrian exposures in the Northeast, the Adirondacks have been least disturbed.

For just a moment, imagine a wrecked automobile in a junkyard. Most of the body of the car is intact, but the bumper and fenders at the front are badly smashed. As you inspect the damage it becomes clear that the car has been in more than one collision. The damage suggests impacts from several directions. Streaks of paint on the smashed parts of the car seem to have come off another vehicle, or vehicles, which were involved in the collisions. That several colors of paint are visible confirms the multiple-collision hypothesis. Now, look more closely. You discover hammer marks, stretch cracks, and fiberglass filler indicating that the car's bumper and fenders have been straightened and patched between collisions. Look carefully at the apparently undamaged parts. Yes, the frame has been knocked askew. Even at the rear of the vehicle you find frame components that have been nudged slightly out of place. If you are a good detective, you put all these clues together and reconstruct an accident record for the car. You decide how many accidents occurred, in what sequence, and with what other vehicles.

And you assess attempts that have been made to repair the damage. You may also decide that the automobile belongs just where you found it—in the junkyard.

If the North American continent can be likened to the car in the junkyard, then the Northeast is the bashed and battered fender. We will attempt here to reconstruct the history of the continent by examining the crumpled, stretched, and rearranged rocks in the Northeast. No other part of the continent has had a more violent history. No other region offers more revealing clues to the continent's geological past.

The Adirondacks, which lie just to the west of the continent's battered margin, are roughly circular; they appear to have been lifted straight up from below, as if Earth's crust were being pushed from beneath by a giant fist. Geologists call this region a dome. The cause behind the uplift is unknown, but it is probably related to thermal forces in Earth's upper mantle. Fortunately, the uplift has brought rocks into view that would otherwise be deeply buried. These rocks show us part of the continent as it was *before* the plate collisions that assembled and crumpled the rocks to the east and south.

In most parts of the United States, the Precambrian igneous and metamorphic rocks that form the most ancient body of our continent are concealed beneath a thin veneer of younger sediments. Only in a broad region of Canada near Hudson Bay, called the shield, is the continent's Precambrian frame widely exposed at the surface; there you can walk upon rocks which are two or three billion years old, rocks which go back to the continent's very origin. In the United States, the Precambrian continental "skeleton" protrudes through the "skin" of younger, mostly sedimentary formations in a very few places, one of which is the Adirondacks.

Not so long ago (geologically speaking) the Adirondack region in New York, like the rest of the state, was covered with much younger sandstone and mudstone layers. When the region began rising, pushed by that mysterious ''fist'' inside the Earth, erosion accelerated. The overlying sedimentary strata were tipped upward and eroded away, eventually exposing the Precambrian igneous and metamorphic rocks that underlie all the eastern and central United States but are elsewhere hidden. The Adirondack rocks are our window on the past.

The Grenville Province

The Adirondack peaks are carved from a strip of very ancient rock several hundred miles wide stretching from Labrador to Mexico. Geologists call this piece of the continent the Grenville Province, after a village in Canada where the rocks were first studied more than a century ago (Figure 3.2). Radiometric dating indicates that the rocks were formed during one or possibly two continental collisions between 1.0 and 1.4 billion years ago. Some of the Grenville formations are highly metamorphosed sedimentary rocks, such as sandstones and limey shales, which were almost certainly laid down on the Precambrian continent's edge. Others were formed from volcanic materials added to the continent at the time of the collisions. The Grenville Province rocks are the oldest evidence of tectonic activity in eastern North America.

Because the events that added the Grenville rocks to North America happened so long ago, they are very difficult to reconstruct. The meager geologic evidence suggests that about 1.4 billion years ago the region experienced continental rifting. The ancestral North American continent was stretched and broken apart. The piece that broke off may have been a slab of continental crust, called the Baltic Shield, which is now found in Scandinavia. (The rocks in the Baltic Shield are of the same age and type as the rocks in the Grenville Province, and paleomagnetic evidence suggests that 1.4 billion years ago the Baltic Shield and North America were at the same latitude.) As North America rifted, an

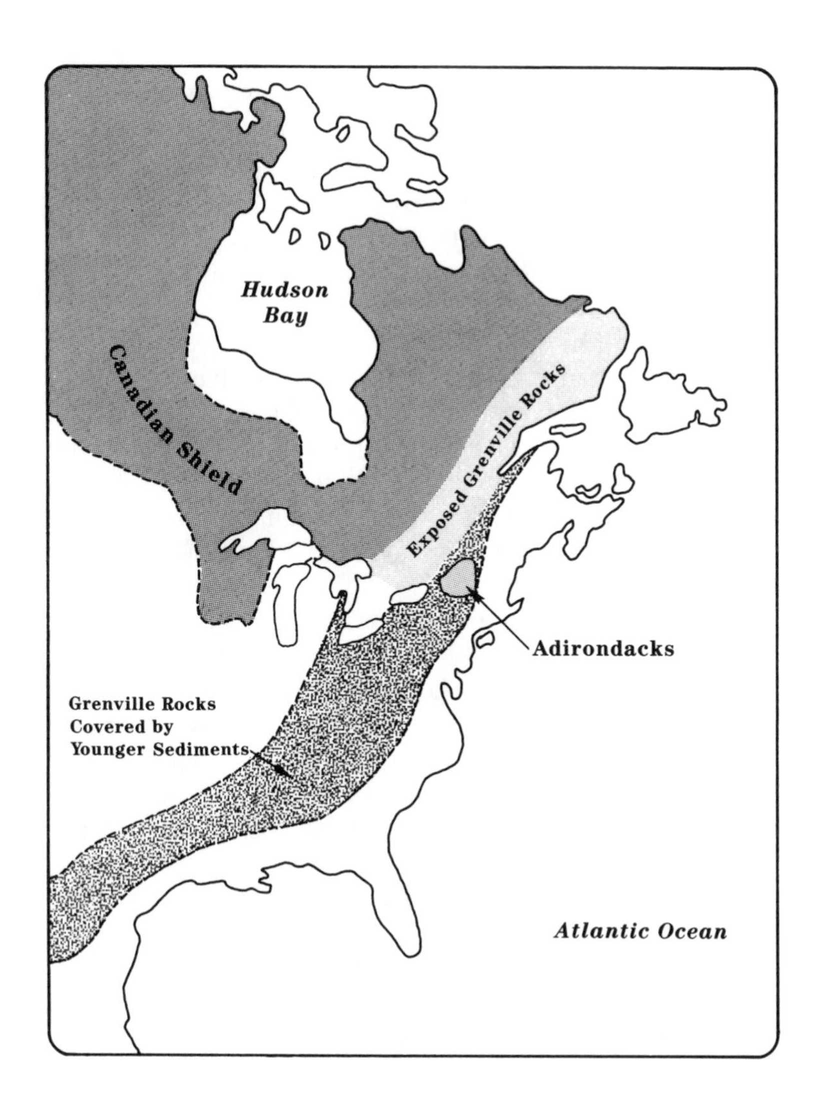

FIGURE 3.2—The Grenville Province is a band of Precambrian rocks near the eastern margin of North America. These rocks are widely exposed in Canada. Over most of the United States, except in the Adirondacks, they are covered by younger sedimentary formations. Still older rocks, more than a billion years old, are exposed in the Canadian Shield.

ocean basin opened between the two slabs of continental crust and thick beds of sediments were deposited along both continental margins. About 1.1 billion years ago, stretch gave way to squeeze—perhaps as the Baltic Shield moved back to briefly collide with North America. Whatever the source of the squeeze, it raised a mountain range that reached from Labrador to Mexico, a towering ridge of snow-capped peaks like the present-day Himalayas or Andes. The rocks in the mountains included sediments that had earlier accumulated on the continental margins and were caught in the crush of plates. The mountains also held volcanic and intrusive igneous rocks, convincing evidence of the heat and pressure that accompanied the converging plates. All this material was squeezed like clay in a fist, twisted and deformed, and forced upward.

Mountain-building events are called *orogenies*. The Grenville orogeny was followed by relative quiet for hundreds of millions of years. As the mountains were gradually eroded away, the crust rebounded, exposing new material to erosion. All told, more than a dozen vertical miles of crust may have been eroded. Rocks that had once lain deep at the root of a mountain range were now exposed at the surface. The Grenville Province had been created.

Sometime before the Precambrian era closed, while the mountain range was being worn down, the Grenville continent was again rifted. A new ocean basin opened to the east. This ocean was not the present Atlantic, but an ocean that is now long vanished, appropriately called Iapetus by geologists. In Greek mythology, Iapetus was the father of Atlas, for whom the Atlantic Ocean is named.

A Gray and (Almost) Barren Shore

The rocks in the Grenville Province were not the first to be added to the North American continent, nor would they be the last. Older belts of crumpled rock have been identified in the exposed shield of central Canada, each standing as mute and murky evidence of even more ancient episodes of continental rifting and collision. But these earlier chapters in the continent's story have little to do with understanding the present geography of the Northeast.

Grenville rocks are prominently exposed in the Adirondacks. Smaller outcroppings can be found in the Berkshire Hills and Green Mountains, in the Hudson Highlands, and in Manhattan. The Hudson River carves a gorge through Grenville rock at Storm King Mountain, near West Point. All these exposures contribute bits and pieces to the picture of what North America looked like toward the end of the Precambrian era, 800 to 600 million years ago. The mountains pushed up in the Grenville orogeny were long gone. North America lay astride the equator, and the Iapetus Ocean washed against a *southward*-facing shore.

On the Grenville coast of that older North American continent, a warm blue-green sea sparkled in the sunlight. Gentle waves lapped the sandy strand. Breezes teased up white clouds in an azure sky. The tropic shore might have been a vacationer's paradise, had anyone been there to enjoy it—but another 600 million years would pass before human beings appeared on the planet. Indeed, not one visible creature inhabited the shore: no palm trees, no dune grasses, no birds. No fish swam in the waters. No insects stitched their music on the humid air. But life was there, abundantly, in the shallow coastal waters, in the tidal mud flats, and in the deeper sea. It was microscopic, mostly single-celled, and invisible. Algae and bacteria were there in staggering numbers. The microbes were the Earth's sole inhabitants.

Unlike animals with shell or bone, soft-bodied microbial organisms do not leave a conspicuous fossil trace in rocks—with one prominent exception. In the Grenville rocks of upper New York state can be found fossil stromatolites. These enduring structures appear in Precambrian rocks around the world, and are among the earliest evidence for life on Earth.

The stromatolites in the Grenville rocks of the Adirondacks are proof that the late Precambrian shore in the Northeast was not devoid of life.

Stromatolites (sometimes called "cabbage heads") are layered rocky domes or cylinders that range in size from baseballs to bushel baskets. They are made by threadlike algae that live in sticky mats in shallow water. The algae require sunlight for photosynthesis. But the sticky algae mats trap particles of carbonate mud and grit from the water that washes across them. This buildup soon obscures light from the sun. The algae must then grow through the carbonate layer and form a new sticky mat on top. This mat becomes gummed with carbonate particles, and again the algae reach upward. And so it goes, layer after layer. In their search for sunshine the microscopic algae build sizable domes and towers (Figure 3.3). Stromatolites are not fossilized living creatures; rather, they are fossil structures that were *created by* living creatures. The geologist sees them as evidence for the past presence of life as vividly as the ruins of an ancient city reveal a human presence to an archeologist. The stromatolite-building algae appear to have had their golden age about two billion years ago. Today, in a few warm-water environments worldwide, they go on about their work, but they are much less common than formerly. Modern stromatolites are strikingly similar to the fossil stromatolites found in rocks billions of years old. The stromatolite-building algae are one of the most enduring branches on the tree of life.

In shallow warm-water bays and inlets on the Grenville shore, stromatolite algae colonies that looked like overturned bushel baskets were changing Earth's face. Life was building enduring structures that would be recognizable a billion years later in rocks that had been lifted, twisted, squeezed, and overturned. Life was busy, and not just forcing its imprint onto the rocks; the stromatolite-building algae contributed something even more vital to the unfolding story of the

FIGURE 3.3—Above, modern stromatolites grow on the coast of western Australia. Below, a fossil stromatolite lies embedded in Precambrian rock.

Earth. They helped create the prominent oxygen component in the atmosphere. *For every breath we take, we owe a debt to the Precambrian microbes.*

Photosynthesis is the chemical process by which living organisms use sunlight to build sugars, life's way of taking energy from the sun. One of the chemical byproducts of photosynthesis is oxygen. During their golden age, the stromatolite-building algae released enormous amounts of oxygen into the atmosphere, increasing its oxygen component from essentially zero to about 20 percent. As the atmosphere changed, life evolved new chemical techniques for using oxygen to advantage. The most remarkable new technique was respiration, using oxygen to release efficiently energy from sugar. Respiration, developed first in single-celled microbes that exchanged gases through their cell membranes, eventually led to the balance between photosynthesizing plants and respiring animals that today maintains a steady proportion of oxygen in our environment.

Earthquake Country

The current uplift of the Adirondack region must surely subject the rocks to stress. The Blue Mountain Lake area, in the heart of the mountains, has intensely interested seismologists at the Lamont-Doherty Geological Observatory in Palisades, New York. The area periodically gives rise to swarms of small unfelt earthquakes and occasionally to a sizable quake that is felt. A swarm with hundreds of microearthquakes in 1971 led to the first successful prediction of a felt earthquake in North America. That quake, which had a magnitude of 2.5 on the Richter Scale (not large enough to cause damage) occurred July 27, 1971. An even larger earthquake, on October 7, 1983, registering 5.2, was felt as far away as Buffalo and Maine. Geologists at Lamont-Doherty are presently at work using earthquake data to trace the active faults in the region.

An occasional earthquake is the price we pay for the

spectacular mountain scenery and recreational environments of the Adirondacks. It is also the price we pay for our window on the past. The uplift that produced the mountains, lakes, and wild valleys also brought into view the long-buried margin of an ancestral North American continent, the "Northeast" a billion years ago. That vanished Grenville shore is the earliest geography of our region that can be reconstructed with any certainty.

The oldest known rocks in the Northeast date from the Precambrian era. These metamorphosed sedimentary rocks, which are now exposed in the Adirondack Mountains, were formed between 1.0 and 1.4 billion years ago during the Grenville orogeny, the earliest known episode of tectonic activity in the Northeast. Within these rocks are fossils of stromatolites, structures built by blue-green algae that lived in Precambrian oceans. Photosynthesis by these microbes produced oxygen, making possible the evolution of animals that use oxygen for respiration.

Chapter Four

The Taconic Upheaval

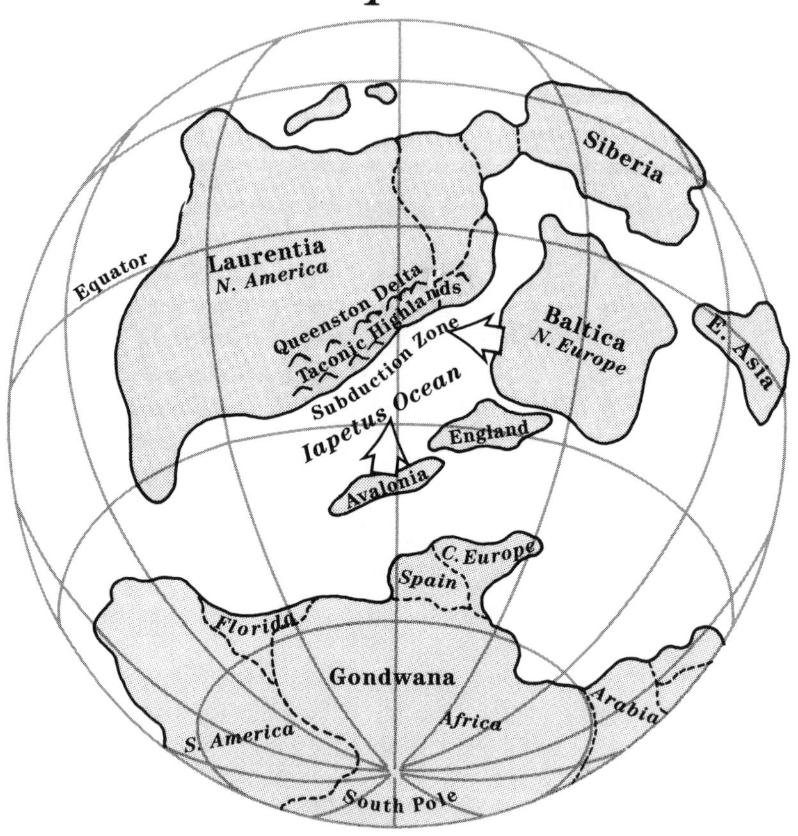

Precambrian	Paleozoic			Mesozoic		Cenozoic

4600 545 445 355 250 145 65 2

Millions of Years Before Present

Scale not proportional

New York City's skyscrapers are clustered within two districts of Manhattan Island—downtown and midtown. In these two parts of the island the bedrock rises to within a few feet of the surface and provides a solid foundation to support the weight of tall buildings. Between midtown and downtown, the bedrock dips hundreds of feet below the surface, and these "basins" are mostly filled with glacial deposits, unconsolidated materials that would make a shaky footing for a skyscraper. Where the island's bedrock foundation is naturally exposed for inspection—or where engineers building foundations, aqueducts, bridge abutments, or subway tunnels have blasted into or through the city's bedrock backbone—the rock is recognized as metamorphic. The New York City bedrock is schist, marble, and gneiss, all ancient sedimentary rocks in highly metamorphosed forms. The source rock for the schist was a sandy mudstone; for the marble it was limestone; and for the gneiss it was sandstone. These rocks were given their present metamorphic character by intense pressure and heating. Wherever on Manhattan Island these rocks are exposed, they show evidence of intense compression and folding.

The Manhattan Island rocks were given their present character in the early Paleozoic Era, about 500 million years ago. Unlike the older, more intensely deformed Grenville rocks, these Paleozoic rocks offer a relatively clear picture of their origin and subsequent transformations. The sediments that ultimately formed these rocks were deposited in low-lying basins on the Grenville shore of the Iapetus Ocean. Sands, carbonates, and sandy muds accumulated for millions of years, became compacted and cemented, and turned to stone. Eventually, Iapetus began to close, squeezing the basin laterally so that the originally horizontal strata were crumpled up into high mountains, perhaps more than 10,000 feet high—alpine, snowcapped peaks. At the mountains' heart, the sedimentary rocks were transformed by heat and pressure into gneiss, marble, and schist. Inevitably, over millions of years, erosion cut the mountains down, down to sea level. Upon the exposed root of such a mountain range New York City is built.

The same crushed and metamorphosed rocks that are found on Manhattan Island can be traced northward all the way into Canada. Most of the region between the Connecticut River and the Hudson River–Lake Champlain valley consists of faulted and deformed rocks of Cambrian and Ordovician age (545–445 million years before the present) that are often highly metamorphosed (Figure 4.1). The Green Mountains, the Berkshire Hills, and the Taconic Mountains in eastern New York are all part of these structures. Another branch of these ancient and highly deformed rocks crosses the Hudson River at Bear Mountain and extends down across northwestern New Jersey into Pennsylvania. The east–west squeeze that twisted and transformed these rocks occurred during the Ordovician period, from about 490 to 445 million years ago. That event is called the Taconic orogeny. It was the first of three great mountain-building thrusts that would shape the Northeast in the Paleozoic.

Before the Appalachians

To understand the present topography in western New England and eastern New York state we have to first reconstruct the landscape as it was before the deformations. In Cambrian times, between 490 and 545 million years ago, the eastern coast of North America was geologically at rest. The mountains produced by much earlier tectonic events, including the Grenville orogeny, had been erased by erosion. The continental margin was a broad, flat plain that declined gently toward the continental slope and the deep Iapetus ocean basin. The Iapetus was then an expanding ocean, with a rift valley down its center, not unlike the present Atlantic. But the boundaries of the Iapetus Ocean were not quite the same as those of the Atlantic; northern Ireland and

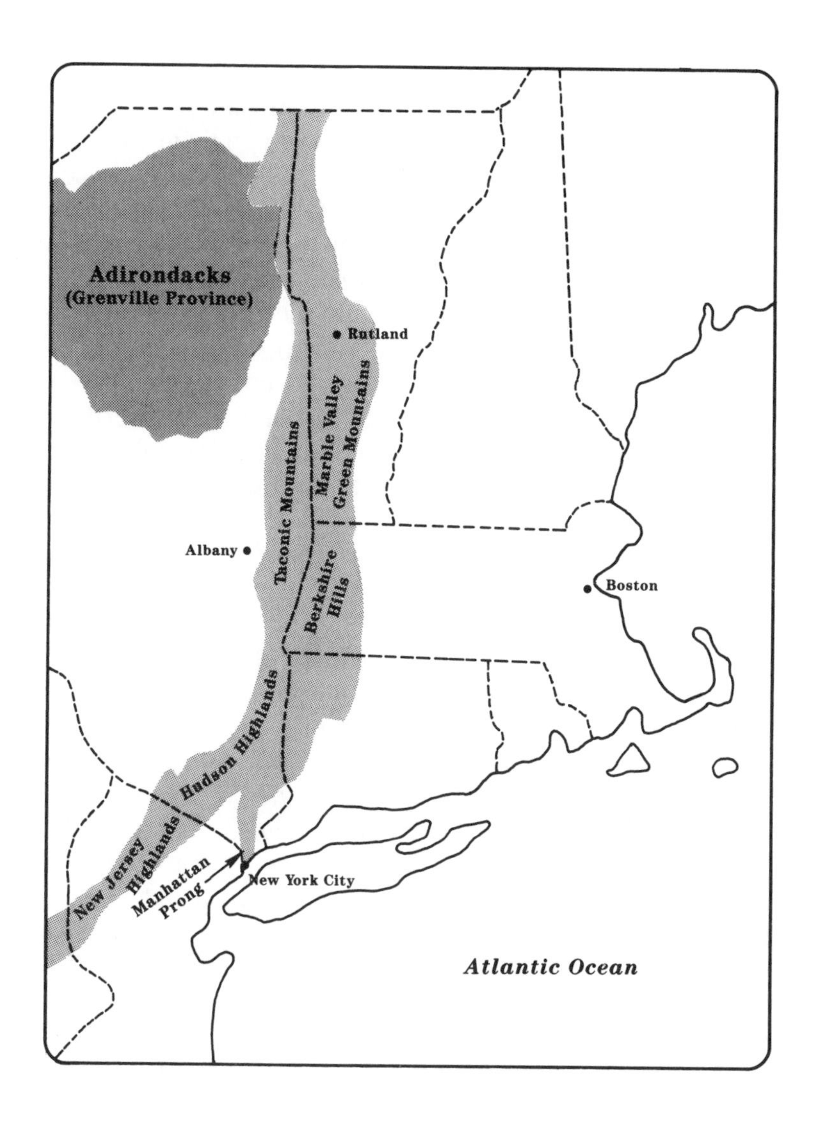

FIGURE 4.1—This band of ancient rocks stretching from eastern Pennsylvania to Canada was affected by the Taconic orogeny, about 500 million years ago.

Scotland were attached to Greenland on the western side of the Iapetus, and southern Ireland and England were part of the European continent in the east. Newfoundland was similarly split into western and eastern halves divided by the widening sea. All of New England that presently lies east of the Berkshires and Green Mountains was not yet a part of North America. (We shall see later how the New England rocks came to be affixed to the continent.) But the edge of that earlier, less bulky continent was not where the Berkshires and Green Mountains stand today; rather, it was many tens of miles farther east. In the first of the three squeezes that were to come, the entire continental margin was compressed, crunched up like the front of an automobile in a collision.

As they eroded, the last of the Grenville highlands contributed sands to the Cambrian continent's gently sloping margin. The continental margin sagged under these sediments and at last the sea intruded upon the land. Sea levels all around the world stood very high in late Cambrian and early Ordovician times. North America in those days was part of a continent geologists call Laurentia, which included Greenland. Most of Laurentia was submerged by a shallow sea, although a broad area in central Canada was dry. The sea that lapped the continent was warm, for at that time North America lay astride the equator, but tipped onto its side, so that the present east coast of the continent faced south; the northeastern "United States" was then at a latitude of about 25 degrees, south of the equator. By mid-Cambrian times the nearby highlands that had supplied erosional sands or muds to the floor of the shallow sea that submerged the coastal plains were gone. Calcium carbonate deposits began to accumulate on the older sands, eventually forming a thick carbonate platform, lightly covered by the sea, which stretched from Newfoundland to Alabama. This carbonate bank consisted of minerals precipitated from sea water and the skeletons of microscopic marine organisms.

With time, the carbonates became limestone, and where those limestones are today exposed we see evidence that the carbonate platform maintained itself at or near sea level. Mud cracks, stromatolites, and other evidence preserved in the rock indicate an intertidal or shallow subtidal environment. The carbonate platform was probably like the present day Bahama Banks, though much larger. The limey deposits accumulated throughout the Cambrian and into Ordovician times. Eventually the carbonate bank became nearly two miles thick, thinning toward the interior of the continent (Figure 4.2a). These carbonates became the limestones and dolomites that presently crop out all across the Appalachian province. Where the limestones were subsequently transformed by metamorphism (during the Taconic orogeny), they provide the low-grade New York City marbles and the splendid commercial marbles of Vermont.

Age of Trilobites

Few regions of eastern North America were dry in Cambrian and early Ordovician times; those few were also lifeless and bare. Not one plant softened the continent's rocky face. Yet life existed in the oceans, where a remarkable revolution in the history of life was unfolding.

For three billion years, life on Earth had consisted solely of microscopic organisms. The bacteria and the algae that had ruled the Earth continued to have a pervasive and profound influence upon the planet. They built rocks, modified the atmosphere, and regulated the climate. But a visitor to Earth from another planet would have required a microscope to discover that the Earth was inhabited. Then, toward the end of Precambrian times, came an explosion of new life forms: variegated, multicellular, visible to the unaided eye. Suddenly the seas were teeming with trilobites, brachiopods, mollusks, echinoderms, nautiloids, and ostracods. The cause or causes behind this spectacular revolution are still debated by evolutionary biologists. The

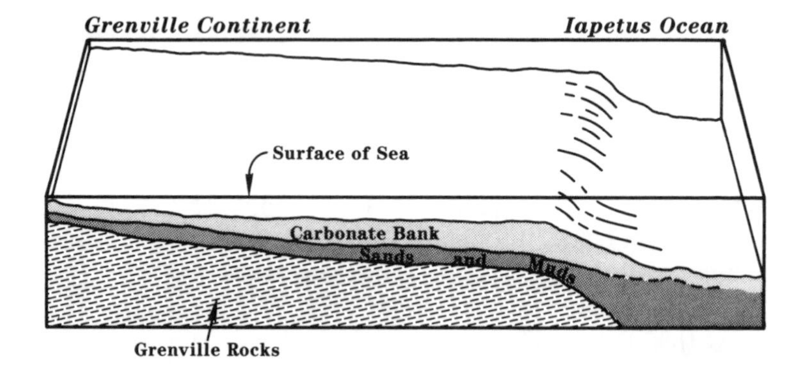

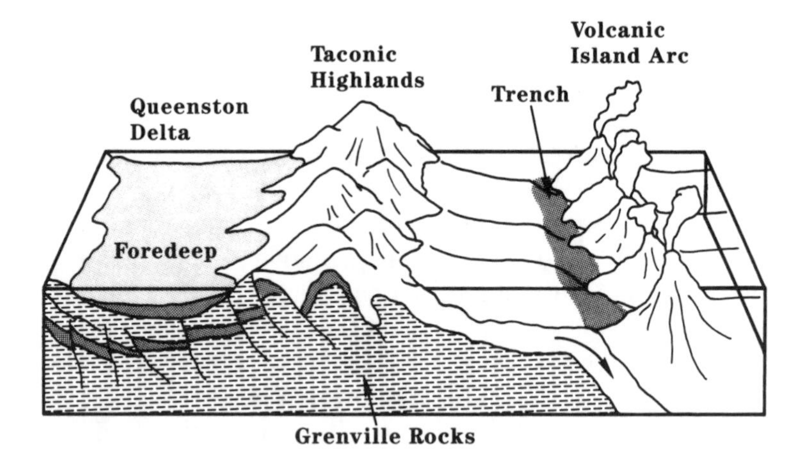

FIGURE 4.2 a—This cross-section shows the carbonate bank deposited on the coast of the Grenville continent. b—Subduction of the floor of the Iapetus Ocean squeezes the margin of the Grenville continent, raising the ancestral Taconic Highlands.

generous energy resources afforded by respiration were certainly instrumental in the revolution. Sexual reproduction may have been even more significant: a new way of producing offspring that utilized two parents and a lively mixing of genes. Variation is the driving force in evolution; with sexual reproduction, genes are shuffled in every generation.

The dominant fossils in Cambrian rocks in the Northeast are trilobites (Figure 4.3). The Geological Highway map published by the American Association of Petroleum Geologists (see Bibliography) tells you where to find them. Trilobites take their name from the three longitudinal segments, or lobes, which compose their bodies. The trilobites were arthropods, invertebrates with segmented bodies and jointed legs; modern arthropods include crabs, lobsters, insects, and spiders. The trilobites' soft bodies were protected by flexible, jointed armor, and growth occurred by molting. Trilobites had mouth parts for chewing small pieces of food, and most had eyes. They crawled over the sea floor and burrowed into sediments. Some trilobites may have been able to swim or float. The smallest trilobites were as big as your thumbnail, and the largest were as long as your forearm. The larger species were the battle cruisers in Cambrian seas, scuttling along the limey or muddy bottoms in search of prey.

Repeated mass extinctions of North American trilobites during late Cambrian times suggest that fluctuations in climate or sea level were common; again and again, numbers of species quite suddenly disappear from the fossil record. Some circumstantial evidence suggests that, after each extinction event, new species expanded their domains to occupy vacated environments. These robust colonizers seem to have come from among the trilobites that lived in deep, cool waters at the continent's edge. Deep-water species would more likely survive a temporary cooling of the sea or a drop in sea level than would shallow-water species. Sug-

FIGURE 4.3—Trilobites roam the floor of a Cambrian sea.

gested agencies that could have suddenly cooled the sea are massive volcanic events or asteroid impacts that blasted huge quantities of dust into the atmosphere, shrouding the planet in darkness. Variations in sea level could have been caused by plate-tectonic movements.

The last Cambrian extinctions eliminated large numbers of trilobite genera. The trilobites' golden age came to an end as the Cambrian closed (490 million years ago). The trilobites' diminishment seems to have cleared the way for rapidly diversifying new life forms during the Ordovician. By late Ordovician times the sea floor was lush with colorful new creatures, including stalked crinoids, reef-building corals, bivalves, snails, and starfish. But even as these organisms diversified and thrived on the Laurentian shore, new tectonic events began to decisively effect their passive shallow-water carbonate-bank environment.

Iapetus Closes

Continents drift across the face of the Earth by fits and starts. The engines that drive the moving plates are presumably heat and gravity within the Earth, which set the deformable (plastic) rocks of the mantle roiling in great convective loops. In recent years, geologists have begun to map these loops by applying computer analysis to seismic (earthquake) waves. The technique resembles the computer-assisted tomography (CAT scan) by which medical researchers nonintrusively map tissue inside the human body. In fact, seismologists owe medical researchers a debt for developing analytical techniques and computer software. The speed at which earthquake waves move through the Earth's interior is sensitive to the temperature and density of the rocks. By analyzing the records of many earthquakes from seismic stations worldwide it is possible to generate computer maps that show regions of rising and sinking material in the mantle. These maps of the Earth's churning interior are extraordinarily complex. It comes as no surprise that surface-plate movements over the long geologic ages are equally intricate.

By mid-Ordovician times the Iapetus Ocean had ceased widening and began closing. A continent called Baltica, which was essentially today's Northern Europe (including Scandinavia, but excluding northern Ireland and Scotland), began to approach Laurentia from the southeast, drifting up from southern latitudes. As the ocean narrowed, the Iapetus ocean floor was gradually subducted along a southeastward-facing trench with a line of volcanos (an island arc) on the seaward side of the trench (Figure 4.2b). Eventually, when all the intervening ocean floor was subducted, the continental margin and the island arc were crushed together. The carbonate bank was lifted, faulted, and telescoped, thrusting limestones and deep-water sediments above sea level, and creating high mountains. These uplifted rocks were the original Taconic Highlands. Here and there, the ancient basement rocks of the older Grenville continent were pushed up through overlying sedimentary and volcanic rock, as a bone is pushed through the skin in a compound fracture of a human limb. And to the west of the mountains, the crust sagged and was covered by the sea. This shallow water-filled basin separating the deformed continental margin from the undeformed interior is called a *foredeep*.

The Queenston Delta

As the magnificent Taconic Highlands rose, erosion cut them down. The debris was carried by flowing water to the east and west, into the offshore subduction trench in the east, and into the foredeep in the west. Sands, gravels, and muds began to build up on the foredeep floor and helped depress it further. Eventually, the rate at which sediments were supplied from the Taconic Highlands exceeded the rate at which the foredeep was subsiding and the basin was filled. Deposition of sediments in a marine environment gradually gave way to the kind of deposition that occurs in river deltas and dry lowlands. A great wedge of sedimentary debris finally extended hundreds of miles westward toward the continental center. Those materials are today referred to as the Queenston Delta.

In general, the Queenston Delta materials are coarser in the east, where they were deposited at the base of the new mountains by fast-moving streams, and become progressively finer as one moves westward, where they were presumably carried by more sluggish water. At the end of Ordovician times and as the Silurian began (about 445 million years ago), the Taconic Highlands were at their grandest, and fast-moving streams deposited coarse debris at the western base of the mountains, where, even as the foredeep was being filled, waves still lapped against Taconic foothills. More than a thousand feet of a rough quartz-pebble conglomerate, the Shawangunk conglomerate, were laid down. This conglomerate consists of sand and gravel eroded from the mountains and deposited along the beach on a wave-battered shoreline. Subsequently lifted, that material now caps the dramatic cliff-girded mountains near Kingston, New York. These conglomerates have also been preserved at Kittatinny Mountain in New Jersey.

Mountains out of Place

While the Queenston Delta was being deposited, the tectonic squeeze from the east remained strong, at times so strong that blocks of uplifted deep-water sedimentary rock were pushed westward from the Taconic Highland onto the newly accumulating foredeep sediments. In this way, blocks of old continental crust found their way onto the top of younger rocks.

One place at which we can see older rocks thrust dramatically over younger formations is at Lone Rock Point on the shore of Lake Champlain, just north of Burlington, Vermont (Figure 4.4). Here is one of the most vivid examples of thrust faulting in the country, a geological showpiece. A

FIGURE 4.4—This thrust fault appears at Lone Rock Point on Lake Champlain. Older carbonate rock is lighter in color than the younger sedimentary formations that lie beneath.

massive block of dolomite, a limestone-like rock that was part of the old carbonate platform, has been pushed over a much younger bed of shale (mudstone) that formed from the erosional debris of the Taconic Highlands.

The vigorous westward thrusting of the rising Taconic range left a far more substantial, if less obvious impression on the Northeast landscape. A long valley today separates the Taconic Mountains in New York from the Green Mountains in Vermont and the Berkshire Hills in Massachusetts. Route 7 follows the valley from Burlington, Vermont, through Rutland, Bennington, Williamstown, Pittsfield, and Great Barrington, to Connecticut. To the east of the valley are mountains of resistant rocks that in late Ordovician times were at the core of the impressive Taconic Highlands. To the west of the valley are the present Taconic Mountains, composed of resistant rocks that lie atop younger limestones and dolomites. For a long time, the Taconic Mountains were a geological puzzle—the rocks in those ridges did not seem consistent with their surroundings. Now we understand that they were thrust by compressive forces tens of miles westward, from the area that is today the Green Mountains onto sediments in the Ordovician foredeep. The Taconic Mountains in New York were contributed to that state by Vermont!

Marble Quarries

The valley that separates the present Taconic and Green Mountains lies where the rocks from the ancient carbonate platform are now exposed at the surface. The carbonate rocks are more easily eroded than the rocks to either side, and so they are worn down, forming the valley.

The ancient limestones and dolomites were cooked and squeezed deep inside the Earth by the tectonic forces that thrust the Taconic rocks skyward and westward. The heat and pressure altered a dull, lusterless gray rock into a sparkling, crystalline substance known as marble. Marble can be found from one end to the other of the region affected by the

Taconic orogeny, from Manhattan to Canada, but it is only in the valley between Dorset and Brandon, Vermont, that the stone is high enough in quality to be suitable for building or statuary. Pure marble is snow white, such as the marvelous Cararra materials carved by Michelangelo. The beautiful colors and patterns in the Vermont marbles are caused by impurities. Many marble quarries can be observed from Route 7. At Proctor, Vermont, the Vermont Marble Company maintains an exhibit explaining the geology, quarrying, and uses of marble.

Mountains Put to Rest

By the end of the Ordovician period the Taconic uplift had ceased. Whether Baltica actually collided with Laurentia along the subduction zone, or whether the two continents merely approached each other without making contact, crushing an arc of volcanic islands against the mainland, is uncertain. Certainly, the Iapetus was greatly narrowed. Fossils of shallow-water marine faunas from the early Ordovician are quite distinct on Laurentia and Baltica, but by late Ordovician times the fossils are very similar. By the time the Taconic Highlands were rising most vigorously, many species of life were apparently able to cross from one side of the Iapetus Ocean to the other.

During the Silurian period of geologic history (445–415 million years ago) the majestic Taconic Highlands were eroded at last to sea level. The wedge of sediments to the west of the diminishing mountains accumulated finer and finer materials, mostly muds carried by sluggish streams over gently undulating terrain. Today, these Silurian deposits are exposed along the Mohawk River valley in eastern New York, and between the New York Thruway and Lake Ontario in the west. Tens of thousands of cubic miles of materials were removed from the Taconic Highlands and deposited as far west as Ontario. At last, the sea returned to cover parts of New England.

In their finest hour, the Taconic Highlands stood as a majestic bastion along the eastern coast of Laurentia, as the Andes Mountains stand today along the subduction zone where the floor of the Pacific pushes against South America. By the time the Silurian period ended, a mountain system that was more than a thousand miles long and perhaps fifteen or twenty thousand feet high had been erased. But it did not vanish completely. It still resides in the late Ordovician and Silurian sedimentary rocks that reach westward from the Hudson Valley to Ontario and Ohio. (Over much of New York and Pennsylvania, the Taconic erosional debris of the Queenston Delta is buried by younger formations.) The Taconic Highlands also left their impression on western New England and eastern New York. The Green Mountains, the Taconic Mountains, and the Berkshire Hills are the recently uplifted and eroded roots from that ancient range. And if you stand on any bedrock outcrop in Central Park in New York City, you are on rock that 440 million years ago was deep inside a mountain range of alpine grandeur.

The Taconic orogeny, the first of three mountain-building events that deformed the Northeast during the Paleozoic era, occurred between 500 and 440 million years ago. This event, probably caused by the Northeast's colliding with an island arc as the Iapetus Ocean began to close, formed the Taconic Highlands. Today the highlands' uplifted sediments contain fossils of trilobites, crinoids, bivalves, and other life forms. The highlands were eroded during the Silurian period, and a large wedge of their erosional debris, the Queenston Delta, was deposited to the west.

Chapter Five

Exotic Terranes, Green Terrain

Precambrian	Paleozoic	Mesozoic	Cenozoic

4600 545 445 355 250 145 65 2

Millions of Years Before Present

Scale not proportional

A hot word in geology these days is "terrane." (The full technical expression is tectonostratigraphic terrane.) A terrane is a block of continental crust with significantly different characteristics from those of the crust immediately adjacent to it. A terrane may have originated in a different part of the globe, at a different time, or by a different process. It came to its present position by riding the moving plates that make up the Earth's dynamic crust. A terrane is a bit of continental crust that has been shuffled by plate tectonics into a place where we'd not expect it.

The terrane idea was formed in the 1970s, when U.S. Geological Survey teams working to resolve land-use disputes in Alaska discovered that the entire state is made up of slabs of continental crust that were crushed together willy-nilly over the past 180 million years. A paleomagnetic study indicated that many crustal slabs had their origin at latitudes far to the south. One well-studied chunk of Alaskan crust, Wrangellia, once lay south of the equator. Riding on a drifting plate, Wrangellia collided with Oregon 70 million years ago. Fragmentation and more drift subsequently carried slivers of Wrangellia as far to the northwest as the Wrangell Mountains in Alaska. The Wrangell Mountains therefore have more in common with some rocks in eastern Oregon than with their immediate neighbors in Alaska.

We now know that the entire western coast of North America is a collage of exotic terranes—oceanic ridges, volcanic islands, chunks of old continents, ocean-floor sediments—that have been plastered onto the growing continent layer by layer as the Earth's surface plates slip and slide. The continent has grown by accretion, like a rolling snowball, adding on layers as it drifted westward. In recent years, geologists in the East have adopted the terrane concept from their western colleagues, and they are reinterpreting the Northeast's geological history. Since the 1960s, the standard story about Northeast geology involved supposed collisions of Europe and Africa with North America in three successive pulses between 500 and 250 million years ago. (We have recounted the first of these pulses, the collision that raised the ancestral Taconic Mountains in eastern New York.) According to this ''traditional'' view, the Northeast rocks were crumpled as Pangaea was gradually assembled from Europe, Africa, and the Americas. When later, in the Mesozoic, Pangaea broke apart and the continents began to drift toward their present positions, bits of Europe and Africa were left behind on the North American continent.

Now the story has become more complicated. Geologists have identified parts of eastern North America that appear to have the same confused relationship to one another as do the terranes in western North America, although the eastern terranes assembled much earlier. Most of New England appears to be a hodgepodge of exotic crustal fragments that were added to the continent in the Paleozoic era like successive layers of icing to a cake. We now recognize that the Precambrian Grenville Province itself may consist of exotic terranes added to North America in the even more distant past.

The terrane concept has changed our way of thinking about continents. We used to imagine the continents drifting about the surface of the earth looking always more or less as they do today. We now recognize that continents are continually being fragmented and rearranged, chunk by chunk, and growing by accretion of new material from the mantle.

Suspect Terranes

We have seen how during late Ordovician times Baltica approached Laurentia as the Iapetus Ocean narrowed. The floor of that ancient sea was subducted, taken back into Earth's mantle. But not entirely. Less dense ocean sediments resisted subduction and were thrust up over the continental margin, to form part of the ancestral Taconic Mountains. During or soon after the Taconic thrust, an additional mélange of oceanic materials, possibly including an island arc, was crushed against the continent's margin. These volcanic and sedimentary rocks of an oceanic origin now form the so-called Dunnage Terrane, a bit of crust that includes all of northern Vermont east of Montpelier, the northernmost tip of New Hampshire, and all of Maine north of Moosehead Lake (Figure 5.1). The rocks in the Dunnage Terrane were greatly metamorphosed as they were plastered against the continent and in the colossal tectonic events that were to follow. Unraveling their history requires skill and imagination.

To the crumpled and growing continental margin was soon added another band of exotic rocks, the Gander Terrane, which includes most of New England's rocks. From the Connecticut coast between Stamford and Old Saybrook, the Gander Terrane reaches up across central Massachusetts, New Hampshire, and Maine. It touches the Atlantic coast between Hampton, New Hampshire, and Boothbay Harbor, Maine. It includes most of central New Brunswick and part of Newfoundland, where at Gander it acquired its name. Southeastern New England and a strip of the ''Down-East'' coast of Maine are excluded from the Gander Terrane, and have, as we shall see, a different story.

The Gander Terrane's origin is highly controversial. It does not seem to have belonged originally to North America. Quite possibly it was part of a facing continental margin, originally separated from North America by an ocean. It might have split off Baltica, racing ahead of that continent to collide early with North America. (Something similar happened when pieces of Africa broke away and collided with Europe in advance of the major collision that is still approaching; Italy is an African terrane that punched early into the Europe's underside.) The Gander Terrane rocks were metamorphosed mainly in Devonian times (415–355 million years ago), when Baltica at last and definitively collided with Laurentia. Huge masses of granite were intruded into the Gander rocks at about that time. These are the

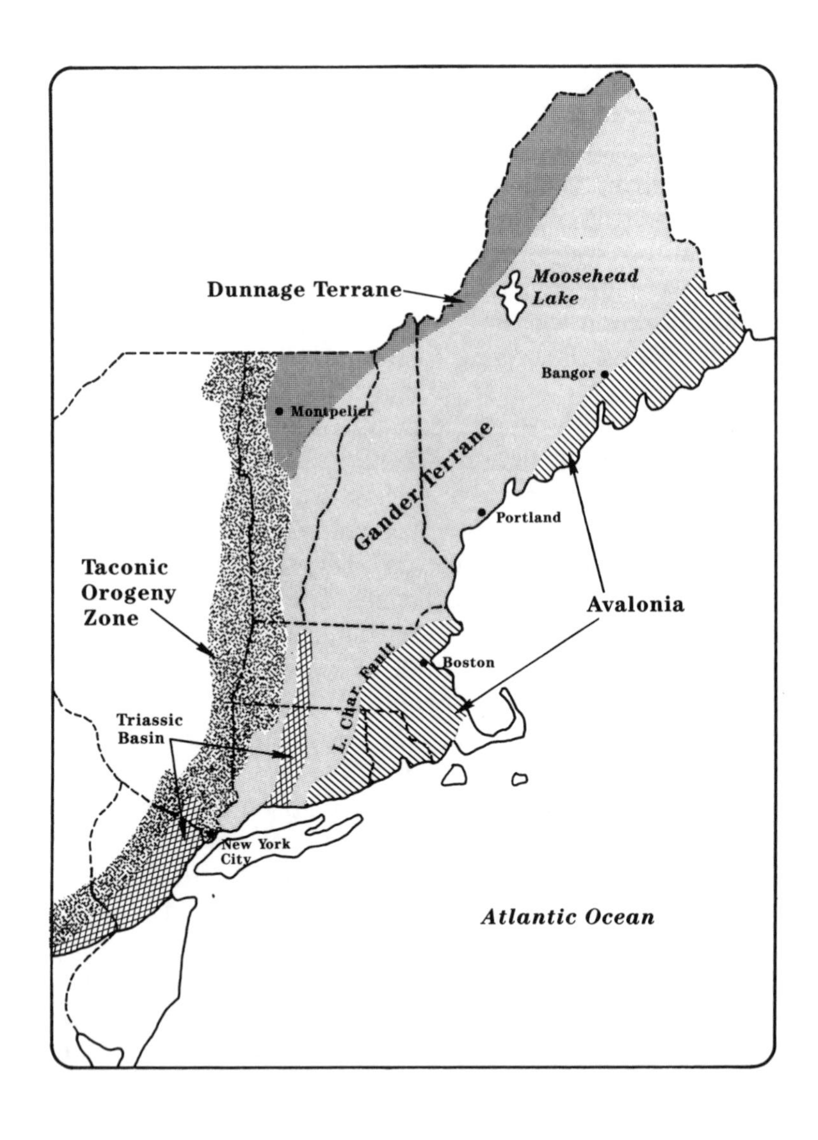

FIGURE 5.1—The terranes of the Northeast east of the Taconic orogeny zone were added to North America as the Iapetus Ocean closed.

granites that give New Hampshire its nickname and supplied the resistant rock that has let Mount Katahdin maintain its lofty dominance over the landscape of Maine.

In the Realm of Paradoxides

As Baltica approached and finally collided with Laurentia, one last terrane was added to New England, probably the most mysterious of all. The terrane is called Avalonia, after the Avalon Peninsula in Newfoundland, which is part of its domain. A strip of the Maine coast south of Bangor, from Boothbay Harbor to the Canadian border, is part of Avalonia. So too is eastern Massachusetts and all of Rhode Island. It is not entirely whimsical to say that Boston is the capital of Avalonia.

Controversial paleomagnetic evidence suggests that Avalonia drifted to its present position from as much as a thousand miles farther south. Other evidence implies affinities between Avalonian and African rocks. We know that Avalonia arrived here from somewhere far off, but no one is quite sure from where. Avalonia may have been a submerged plateau in the Iapetus Ocean, caught in the crunch when Baltica and Laurentia collided. Boston's slab of Avalonia was plastered onto North America about 400 million years ago, when North America itself was still south of the equator. Since that time, the Avalonian rocks have been crunched and shattered, intruded by volcanics and overlaid by sediments, but they have retained an identity that is entirely different from that of their neighboring rocks to the west. Part of the western boundary of Avalonia is the Lake Chargoggagoggmanchauggagoggchaubunagungamaugg Fault, a geological feature with a name as exotic as the terrane it defines. The fault takes its name from a lake in Massachusetts near the Connecticut border; the Indian name of the lake means "You fish on your side, I fish on my side, nobody fishes in the middle—no trouble."

Trilobite fossils found in outcrops at Braintree and

Quincy, Massachusetts, are also found in the Carolinas, the Canadian Maritime Provinces, southern Ireland, and parts of England and Wales—but nowhere else on Earth. These places are all part of one suspect terrane, once presumably united, now broken and dispersed. The temptation is powerful to think of these dispersed Avalonian fragments in the same way as those slivers of Wrangellia scattered up and down the mountain ranges in the West. The Braintree trilobites are of a species known as *Paradoxides*. Paradoxides is indeed a paradox—an exotic, suspect traveler that came with Avalonia from an uncertain far-off place. Paradoxides was one of the biggest Cambrian trilobites, reaching several feet in length. The metamorphosed slab of sea floor which was that trilobite's domain docked against Massachusetts in Devonian times, and it has been here ever since. Where it came from, no one knows.

The Acadian Orogeny

When at last Baltica and Laurentia collided, Avalonia was caught in the crunch, like an automobile caught in a head-on collision between two freight trains. Even as the collision was occurring, the continents were drifting closer to their present positions on the globe.

North America (Laurentia) was slipping northward across the equator, and pivoting into the more familiar north-south alignment. Northern Europe (Baltica) was drifting even more rapidly from far southern latitudes, and moving into position to dock with Laurentia. Siberia, and the flotilla of continental fragments that would become the rest of Asia, were drifting westward toward eventual collisions that would assemble that continent; when Siberia at last collided with Europe it would raise the Ural Mountains. Meanwhile, the southern part of the globe was dominated by a great supercontinent, Gondwana, which embraced the present land masses of South America, Africa, India, Australia, and Antarctica. Florida, too, was at that time part of Gondwana.

When, in the Northeast, the decisive crunch came, and Baltica and Laurentia collided, the exotic terranes that had already docked with Laurentia were crumpled and metamorphosed. Masses of molten material were injected into the debris. A towering mountain range was again raised across the Northeast, and indeed all along the suture between the two continents. Elsewhere, northern and southern Ireland were brought together from opposite sides of the ocean. Scotland was stitched to England, and Scandinavia to Greenland. This great mountain-building impulse is called the Acadian orogeny in North America, and the Caledonian orogeny in Europe. The collision seems to have begun in the north and spread southward. Scandinavia and Greenland were in contact by Silurian times (445–415 million years ago); the crunch that affected the Appalachians came later, in the mid-Devonian (about 380 million years ago).

It requires a great stretch of imagination to visualize the splendid mountain range that must then have stood all across the Northeast. The present mountains in our region are only the eroded roots from those earlier ranges, pale intimations of their former grandeur. The granite of many New Hampshire peaks was then a molten mass miles underground, at the root of the Acadian mountains. In that fiery subterranean furnace, liquid rock was squeezed into every fissure of older sedimentary and volcanic formations, reinforcing them, baking them, and giving them the resistant character that would help them stand up to later erosion.

But even as the mountains went up, erosion began its inevitable work, cutting them down. As earlier during the Taconic orogeny, a wedge of erosional debris was deposited at the foot of the mountains. To the west of the Acadian Mountains, this wedge of sediments is called the Catskill Delta, and it extended far across New York and Pennsylvania, burying the older sediments of the Queenston Delta, and pushing westward the edge of a sea that had returned to cover the Taconic Highlands' degraded remnants. Perhaps "delta" is not the right word for these formations. They do contain true delta deposits, laid down where rivers met the sea. But they also include sediments deposited by braided streams in highland valleys, and meandering river deposits that formed farther downstream. There are tidal-flat deposits, swamp deposits, and sediments that were laid down on offshore sand flats and muddy bottoms. These Devonian sediments today cover the whole of southern New York, from the Catskills to Lake Erie. The combined thickness of the Devonian strata is more than 4,000 feet—vivid testimony to the majestic Acadian peaks that contributed this huge mass of eroded material. Kaatterskill Falls, near Catskill, New York, tumbles down across 260 feet of these sediments (see chapter 2). The bulk of this material consists of sandstones and mudstones piled layer upon layer. Some of these strata abound with fossils of marine organisms that lived in the shallow seas that received the sediments from the east. Devonian fishes have been found in considerable numbers across the state; the Devonian is sometimes called the golden age of fishes. A careful searcher would have little trouble finding fossils of sponges, corals, starfish, brachiopods, and mollusks (Figure 5.2). And one can find fossils of land plants—land plants in abundance!

Invading the Land

With Laurentia sutured to Baltica a new continent was formed, called the Old Red Sandstone continent after well-studied British formations dating from that time. Our region lay near the southern shore of this land mass, near the equator. Much of the continent was covered by a shallow sea in which carbonate and reef deposits were common. The unsubmerged part of the Old Red Sandstone continent included land that is now northern Europe, Greenland, and northern Canada. A few large islands trailed off to the west where now lies the central United States. The mountain range thrown up by the Acadian orogeny bisected the Old

FIGURE 5.2—Life in the Devonian sea was dominated by fishes.

Red Sandstone continent, along the suture between Laurentia and Baltica. These mountains furnished the red sediments (rich in oxidized iron) that give the continent its name. Approximately equal volumes of eroded sediments were carried to either side of these highlands—to the Catskill Delta in the west and to the British sedimentary basins in the east. And in these sediments is fossil evidence for a spectacular event—the greening of the continents.

For the first four billion years in its history, the Earth's rocky crust was devoid of life in conspicuous forms. By the early Paleozoic, some terrestrial environments may have been inhabited by algae. Fossil burrows made by tiny wormlike creatures have been found in late Ordovician formations in central Pennsylvania, in rocks formed by consolidated dry soil. Throughout the Ordovician and Silurian, more imposing forms of life began to encroach upon the shore. But we would have seen no swamps, no forests, no grasslands, no conspicuous animals—only vast expanses of bare rock and dusty desert.

Before life could mount a serious invasion of dry land, plants had to evolve new strategies for survival out of water. Without water to support them, upright plants needed rigid stalks and stems. Root systems were required for the plants to anchor themselves, and to procure water and nutrients from the soil. Leaves were necessary to extract energy from sunlight. The first plants to leave the sea lacked many of these features. They did not have roots or vascular systems. They consisted of little more than rigid stems, partly buried in the ground, and probably lived in semiaquatic environments. By late Silurian times, some plants had evolved vascular systems for transporting nutrients and manufactured foods, and spore systems for reproduction. These innovations were dazzlingly successful. Throughout the Devonian period, as the Old Red Sandstone continent was being assembled, plants decisively stormed ashore.

At the turn of the last century, one of the oldest forests

on Earth was discovered near the town of Gilboa, New York, about fifty miles west of Albany. The petrified middle-Devonian "Gilboa forest" has stumps, pieces of flattened trunks, and fronds from fernlike plants. Trees in the forest grew to forty feet high, and some, precursors of the conifers, may have been even taller. Growing among the trees were many other nonflowering plants: fernlike plants, and plants resembling modern club mosses and horsetails, these latter often growing to far greater size than their relatives in the modern era. The Gilboa fossils suggest a tropical forest growing in swampy terrain in the western foothills of the Acadian Highlands (Figure 5.3).

Did animals live in this tropical forest? The Gilboa deposits have yielded the largest record of terrestrial animal life yet found in American Devonian sediments, including centipedes, mites, and spiders. Elsewhere in the world, strata from the same age preserve the fossils of flightless insects and scorpions. The buzz and hum of animal life could indeed be heard in the giant ferny forest. Then, by late Devonian times (370 million years ago), appear the fossils of the first land vertebrates, fishlike animals that waddled ashore on finny flippers. These creatures left the water only of necessity, flipflopping across the muddy banks after food or a better water hole. The journey was almost certainly perilous. We can imagine these lobe-finned fishes gulping air and dragging themselves laboriously up from the water with fins better suited to swimming than walking. But the skills these early amphibians acquired over millions of years were not lost. Fins became feet, lungs evolved, skeletons were modified for life on land. By the Devonian's end, amphibians were scampering among the trees, horsetails, and ferns. A taste for insects may have encouraged them to seek a life on shore.

The future of life was being indelibly written on the Old Red Sandstone continent, in successful experiments that let life move from the placental oceans into the air. Dry winds

FIGURE 5.3—In this view of a Devonian shore, plants have invaded the land, as have centipedes, insects, spiders, and the first amphibious vertebrates.

blew down the western side of the Acadian Mountains and rustled the green fronds in a lush forest. Forest plants extended their roots into the gritty soil and enriched it with their own substance. Shade from leaves and fronds created new terrestrial environments. The Devonian was a time of softening. The continents were greened. Breezes carried new and exotic vegetable odors. Sunlight dappled forest glades with moving light and shadow. Where plants pioneered, animals were not far behind.

Then, about the time these animals were colonizing the land, occurred one of the most devastating mass extinctions in the history of life. Huge numbers of creatures suddenly and mysteriously became extinct. Warm-water marine species were most severely affected by the trauma; the new vascular plants were affected hardly at all. This was only one among several mass extinctions that have affected life on Earth.

During the Devonian period, as the Iapetus Ocean continued to close, the Dunnage, Gander, and Avalon terranes were added to the Northeast. These small land masses were later deformed and metamorphosed when, about 380 million years ago, Baltica (northern Europe) collided with Laurentia (North America). This collision resulted in the Acadian orogeny and the formation of the Old Red Sandstone continent. The erosion of the Acadian mountains built up a second great wedge of sediments, the Catskill Delta, within which can be found fossils of fish, which dominated Devonian seas, and of the first land plants.

Chapter Six

An African Affinity

Precambrian	Paleozoic			Mesozoic		Cenozoic
4600	545	445	355	250	145	65

2

Millions of Years Before Present

Scale not proportional

During the Paleozoic era, drifting continents delivered eastern North America a one-two-three punch. The first and second punches, the Taconic and Acadian orogenies, went to the shoulders and head—they were directed primarily to New England and Maritime Canada. The third punch, called the Alleghenian orogeny, came from below—to the solar plexus. It gave the southern Appalachians their final form.

As the Carboniferous period began 355 million years ago, North America, now part of the Old Red Sandstone continent, which included Greenland and Europe, was still near the equator. A disjointed Asia, consisting of three major rafts of continental crust and several smaller segments, was slowly approaching the Old Red Sandstone continent from the north and east. But the imminent and climactic action was approaching from the south, as the massive supercontinent Gondwana drifted inexorably northward. The Old Red Sandstone continent and Gondwana were separated by a narrowing east–west seaway. In the far southern latitudes of Gondwana, glaciers covered huge land areas, but nearer to the equator the climate was mild. Warm-water continental seas and forested swamps were widespread in the equatorial regions on both the Old Red Sandstone continent and Gondwana.

In the early Carboniferous period, much of the present United States was covered by a shallow sea. On the floor of that warm-water sea the crinoids (sea lilies) experienced great diversification. The crinoids were a stemmed invertebrate with a root system that was attached to the sea floor. At the stem top a cuplike "flower" with five branching arms filtered food from the water and carried it to a mouth at the base of the arms (Figure 6.1). The crinoids looked like plants but were in fact animals. They thrived in vast colonies on the sea floor, weaving and waving in the warm currents like a grassy meadow in the wind. They extracted carbonates from sea water and secreted hard circular plates, stacked like poker chips and connected by tissue, which

FIGURE 6.1—A "garden" of crinoids proliferates on the floor of a Carboniferous sea.

made up the animal's stem and feeding arms. When a crinoid died, these carbonate disks fell to the sea floor and contributed first to the ''crinoidal'' sands that banked the reef communities and ultimately to the extensive areas of continental limestone. In parts of the American West, beds of almost pure crinoidal limestone are a thousand feet thick. The number of crinoids required to supply such enormous quantities of their skeletal parts is staggering. In the east, the crinoid sea intruded into Pennsylvania, across the Catskill Delta sediments. Here, the limey deposits were muddied by erosional debris carried into the shallow sea from the diminishing Acadian Highlands to the east.

The Coal Swamps

The period we are considering might appropriately have been called the Age of Crinoids, but the land plants gave the Carboniferous period its name. Huge swampy forest flourished just west of New England, where the Acadian Highlands met the inland sea. The dominant trees were lycopods, an early spore-bearing plant related to the modern club mosses. Unlike the tiny ground-hugging club mosses of our own era, the late Paleozoic lycopods sometimes grew to be one hundred feet tall and three feet across at the base. Beneath the green canopy of this towering forest was a lush undergrowth of ferns, spreading fine-fingered fronds to catch the shimmering dollops of sunlight that filtered down from above. As these plants died and fell to the forest floor, thick layers of organic materials accumulated, including great numbers of fallen tree trunks. This organic material, carrying the element carbon in abundance, was subsequently buried and processed by time to become the economically important coal beds of Pennsylvania and the Appalachian Plateau. It takes several cubic feet of swamp wood to make one cubic foot of coal. Every Appalachian coal seam is the residue from a forested equatorial swamp that flourished 300 million years ago. Similar forests grew on the eastern slopes of the Acadian Highlands, in what are now Britain and Europe; the organic substance in these ancient swamps became the fuel for the Industrial Revolution in the late-eighteenth and early nineteenth century. England's rise to prominence as an industrial nation, and later to mastery of a world empire, had its beginning in the Carboniferous forests that grew on the European side of the Acadian Highlands.

Imagine a walk through a dark, damp Carboniferous forest. Bring hip boots, because the forest floor is wet, as in today's Great Dismal Swamp or the Okefenokee Swamp on the United States eastern seaboard. A wet environment was essential for the spore-bearing plant's life cycle (more on this subject later). As you walk, on every side, almost choking progress, are towering trees, unlike any living today. The tree trunks are richly textured by the scars of discarded leaves, which give the trunks a rasplike surface. At almost every step you must climb over a fallen log or push aside fern fronds, heavy with moisture. In these late-Paleozoic swamps, insects have finally come into their own. Dragonflies, some with wingspans of more than a foot, and mayflies dart among the fronds. Cockroaches creep on fallen logs. Like the dragonflies, the cockroaches are among the few creatures from the late-Paleozoic era that have survived almost unchanged into modern times to become ''living fossils.'' Take care in your step for amphibians! By now the amphibians had perfected the apparatus—gills to lungs, fins to limbs—for a transition from sea to shore. The late Paleozoic era was the amphibians' time of glory. Compared to their modern cousins the frogs, toads, and salamanders, some Carboniferous amphibians were huge and frightening creatures. Among the most impressive of these were the labyrinthodonts, animals that take their name from the labyrinthine wrinkling and folding of their tooth enamel (Figure 6.2). The fat, shortlegged, inelegant labyrinthodonts wallowed alligatorlike on the watery floors of the swamps,

FIGURE 6.2—In a Carboniferous swamp-forest at the foot of the Acadian Highlands, the hulking amphibian labyrinthodont captures a meal.

basking in sunlight when they could find it or slinking through shadowy water, preying upon fish, insects, and possibly even each other.

How far the forests marched up the western slopes of the Acadian highlands is hard to say. Almost no late-Paleozoic fossiliferous deposits are found in rocks east of the Hudson River. By late Carboniferous times New England had long been upland country, and it would rise higher as Gondwana pushed from the south. Any fossil formations deposited in the Carboniferous were eroded away. Only in the Narragansett Basin, south of Boston, do we find extensive sedimentary deposits suggesting the New England landscape as it was 300 million years ago. Modest coal deposits in southeastern Massachusetts and Rhode Island formed from plant matter deposited in swamps near a river delta. The coal seams in the basin have been mined occasionally but are not extensive enough to be economically viable as a source of energy. Volcanic rocks scattered in the Narragansett Basin suggest that the river delta was a place of lively tectonic activity. These volcanics are evidence that something big was going on, something that would soon give the rocks in eastern North America their final crushing blow.

The Alleghenian Orogeny

As the time of the great coal swamps was coming to an end, about 290 million years ago, the collision between Gondwana and the Old Red Sandstone continent (North America, Greenland, and Europe) was reaching a climax. Gondwana drifted northward, closing the narrow sea that separated the northern and southern continents. The floor of that sea had to go somewhere, and, as always when continents converge, it was subducted beneath one or both of the approaching land masses. Above the diving sea floor stood a line of volcanos.

In the Boston Basin is manifest evidence left by the volcanos that stood inland from the subducting plate margin. In tiny Dane Park, across from Beaver Country Day School in

Brookline, one can find volcanic basalt and tuff deposits. The basalt may have solidified as dikes of lava that pushed through cracks in the crust, or possibly solidified in a volcano's throat. Tuff is formed from ash particles which are ejected from volcanos and which either fall directly onto the ground or flow furiously down a volcano's flank as a cloud of hot gases and ash. Some of the tuff in Dane Park holds "bombs," rock fragments blown explosively from a volcano's mouth. It is interesting to think that today's peaceful, residential suburb once was the venue for such violent activity.

A new range of high mountains was pushed up by pressure from the subducting ocean plate, just as the subducting floor of the Mediterranean Sea has raised the mountains in southern Europe. Hints of the ancient mountain range that then stood all across New England are not hard to find. At several places in the Boston Basin, such as the Arnold Arboretum, are rock outcrops called tillite. Tillite is formed by the consolidation of gravel, pebbles, and sand deposited by melting glacial ice. Because the Boston Basin was near the equator when the tillite formed, we must assume that the source of the tillite was a mountain glacier rather than continental ice sheets like those that blanket Greenland and Antarctica today. Even more striking evidence for the high mountains that then stood in southern New England can be seen where Routes 128 and 28 intersect just south of Boston. On the south side of the Blue Hills, and easily visible to the highway traveler, is a sedimentary rock formation (called puddingstone or conglomerate) holding many large rounded boulders. These coarse sediments could have been deposited only by fast-flowing mountain streams that roared down high glacier-capped peaks, tumbling along boulders as if they were grains of sand. On the valley floors, the boulders piled up in thick layers and were buried, to become—at a later, quieter period in the Northeast's history—a curious roadside outcrop at the foot of the Blue Hills.

When the narrowing ocean floor was wholly subducted,

Gondwana and the Old Red Sandstone continent came into contact. Because continental rock resists subduction, it had nowhere to go but up. The colliding continental margins crumpled and rose.

Most geologists now believe it was the northwestern bulge of Africa that crunched against North America's eastern margin and the underside of Europe. This mountain-building episode, in North America called the Alleghenian orogeny, was responsible for raising the southern and central Appalachians. Across southern Europe, the corresponding tectonic events are known as the Hercynian or Variscan

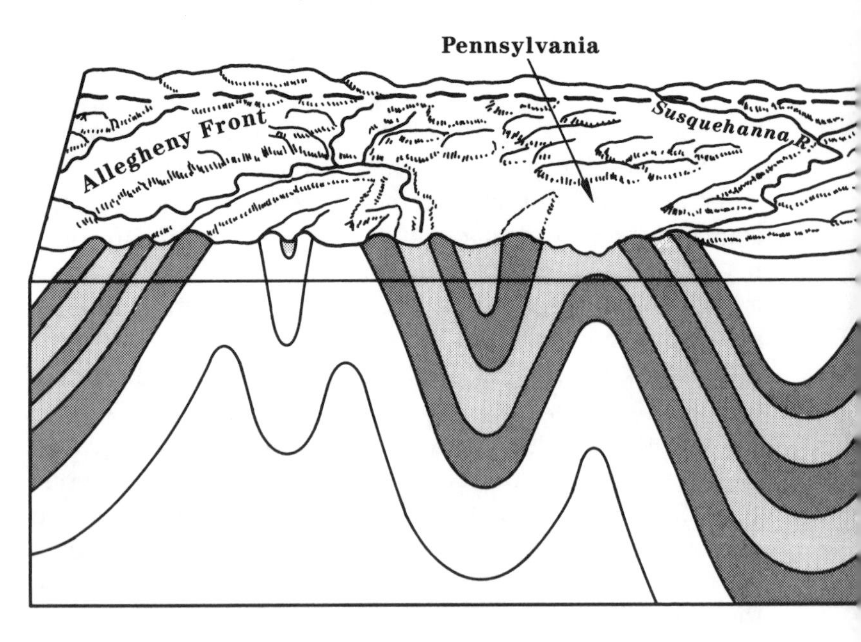

FIGURE 6.3—This cross-sectional view depicts the Ridge and Valley Province of western New Jersey and Pennsylvania. Because the strata

orogeny. In Africa, the crumpling raised the Mauritanide Mountains that bound the Sahara Desert on the north. South America, which was then firmly affixed to Africa, delivered North America a simultaneous blow from down under, along today's Gulf Coast. A continuous mountain range marked the suture, which reached from New England to Texas. The roots of the southern part of this great range, the connecting link between the southern Appalachians and the Ozarks, now are buried beneath the recent sediments of the Gulf Coastal Plain.

Today's southern Appalachians are the remnants of the

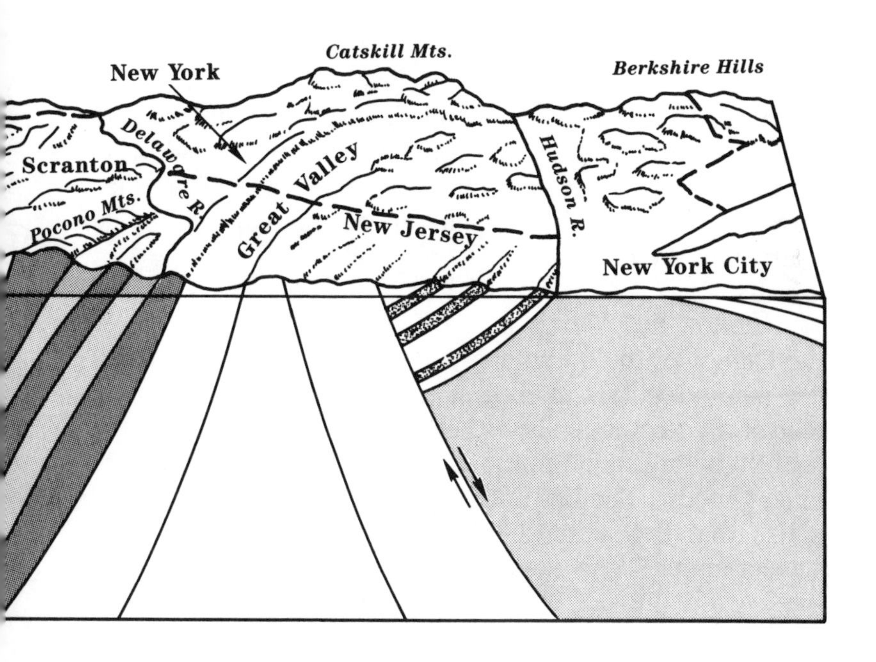

are folded, different types of rock, with different resistances to erosion, are alternately exposed at the surface.

towering snowcapped range that stood along the continental suture in late Paleozoic times. When the continents joined, North America's eastern margin folded up like an accordion. Before the collision, the rocks of Philadelphia were 200 miles farther away from the rocks of Pittsburgh than they are today. The Ridge and Valley Province in Pennsylvania manifests these folds: Where the present erosional surface intersects resistant quartzite rocks, ridges stand; and where the surface intersects easily eroded carbonate rocks, valleys have been cut (Figure 6.3). As you drive across Pennsylvania today, your journey, as revealed by the rocks in the road cuts, takes you backward and forward in geologic time.

Meanwhile, far to the east, Siberia was approaching Europe. The ensuing collision raised the Ural Mountains in central Russia and completed the assembly of an all-embracing supercontinent, Pangaea. This is the supercontinent whose existence was first proposed by Alfred Wegener. By the time the Paleozoic era ended, when the bumping and crunching were finished, all of today's continents had been joined into one land mass that reached from pole to pole, surrounded by one all-embracing ocean that covered three-quarters of the globe.

Hiatus, but Not Quiet

Except for the Narragansett Basin in Rhode Island and a few nearby pockets of rock in the Boston area, a geologic map of the Northeast shows not one outcrop of sedimentary rock from the Late Paleozoic era (the Carboniferous and Permian periods). The key that goes with the geology map explains that this absence of sedimentary rocks signals a "major hiatus." The word "hiatus" makes it sound as if not much was going on, but all it really signifies is a temporal gap in the sedimentary-rock record. A hiatus can occur in a sedimentary-rock sequence because strata of a specific age were never deposited, or because strata were deposited but then eroded before they could be buried by younger sediments. The erosion surface or interval of nondeposition that separates younger and older strata is known as an unconformity. In the northeastern United States, the hiatus in the rock record was probably left by vigorous erosion in the actively uplifting mountain ranges.

The few outcrops of rocks that we can find from this time suggest that the Northeast had high mountains (some of them volcanic) and rushing streams. The mountain flanks were covered by ferny forests that stepped down to lowland swamps in the west and east. Other than that, the geologic record of the Northeast says very little about conditions in the late Paleozoic. With sedimentary rocks so scarce, the fossil record of life at this time is virtually nonexistent.

One can, however, surmise a bit about life in the Northeast from the fossil record elsewhere on the globe. By Permian times (290–250 million years ago) the northern and southern continents were completely sutured. The mountains created by the Alleghenian orogeny were in the middle of the assembled land mass, far from any sea, and close to the equator. To the south of the mountains, on land that is now Africa and South America, plant life was dominated by the so-called *Glossopteris* flora, named for a Gondwana fossil seed fern. *Glossopteris* means ''tongue leaf''; the tongue-shaped leaves grew in clusters near the top of a treelike trunk. North of the mountains grew a very different flora dominated by the lycopods *Lepidodendron* and *Sigillaria*. The mountains separated two distinctive plant communities that had developed in isolation and now, as Pangaea was assembled from its components, found themselves in potential competition.

But something rather different actually happened, affecting both the northern and southern plant communities. As the Permian period progressed, the spore-bearing plants that had dominated the Carboniferous coal-forming forests began to die out. Many plants of that time are today represented only by small ''living fossils,'' such as the horsetails, or the running ground pine that so many people in the Northeast (unfortunately) pull up from the forest floor to make Christmas decorations. The spore-bearing plants were rapidly replaced by the gymnosperms (''naked-seed'' plants), which included in particular the conifers. By the end of the Permian, the Northeast highlands were forested with conifers, even as they are today.

Because diversification among the gymnosperms was one of the most dramatic developments in plant evolution, we might well examine in more detail the difference between seed-bearing and spore-bearing plants.

Spore-bearing plants: As for most contemporary nonmicroscopic organisms, the cells of a typical spore-bearing plant carry two sets of chromosomes (the threadlike structures in the cell that hold the genetic material). The adult plant produces spores with only one set of chromosomes. The spores fall to the ground and grow into small, inconspicuous plants that generate gametes, the male and female reproductive cells that must fuse before they can develop into a new organism. In order to fuse, the male gametes (sperm) must travel through moisture on the surface of the gamete-bearing plant to the female gamete (egg). Successful fertilization forms a juvenile plant with two sets of chromosomes. The juvenile plant grows into an adult, and the gamete-bearing plant dies. Moisture is essential for the spore-bearing plant's reproductive cycle, to allow the sperm to travel to the egg.

Seed-bearing plants: Unlike the ferns and lycopods, seed plants do not release their spores into the environment. Instead, the gametes are produced and fused *within* the adult plant of the spore-bearing generation. This union produces a seed with two sets of chromosomes. The seed is the earliest form of the next spore-bearing generation. The seed is small and well protected and can await favorable conditions before growing into a seedling. A moist environment is not necessary for reproduction.

The hardy seed plants were better able to survive under variable conditions of heat and cold, wet and dry. A warmer, drier desertlike climate prevailed across the supercontinent of Pangaea. This was apparently the stimulus the seed plants needed to begin a rapid diversification. By the end of the Permian period, seed plants had taken over from the spore-bearing plants as the dominant flora, and have continued in abundance even to this day.

At the same time, animals were perfecting new strategies for life on an increasingly arid Pangaea. Amphibians had successfully made the transition from sea to land, but one weakness remained in their pattern of life—the need to return to water to lay their eggs. The reptiles, which evolved from the amphibians, overcame this limitation by developing a hard-shelled egg that could survive out of water. As with the seed-bearing plants, a new strategy for reproduction led to rapid diversification and eventual dominance of reptiles over amphibians.

An Era Ends

Many evolutionary innovations occurred in the late Paleozoic, including seeds and eggshells, but the era ended with disaster—the greatest mass extinctions since life on Earth began. This traumatic event is sometimes called the Time of the Great Dying. Nearly half the plant and animal species became extinct! Seventy-five percent of the amphibians and 80 percent of the reptiles were wiped out. Worst hit were creatures that lived in the sea, especially the invertebrates. Corals, crinoids, blastoids, ammonoids, brachiopods, bryozoans, molluscs, and fishes were devastated. The trilobites, which had suffered several earlier crises, were finally pushed into oblivion. So abrupt and dramatic was this episode of extinction that the sharp break in the fossil record has long marked the division between the Paleozoic (''Old Life'') and Mesozoic (''Middle Life'') eras. What caused the extinctions? The reason is unknown, but debate among paleontologists and geologists has been lively in recent years.

One possible cause might have been the assembly of Pangaea itself. With the supercontinent's suturing completed, very little rifting was taking place in the Earth's crust. Because few rift ridges stood on the floor of the one great ocean, the volume of the ocean *basin* was increased, and sea level fell relative to the continent. The shallow seas that had stood upon the continental platforms drained into the deep ocean basin, leaving shallow-water environments only at the very margins of Pangaea. Overcrowding in these narrowed shallow-water environments may have put a fatal stress on the continental-shelf communities.

The milder, drier climate prevailing on Pangaea may also have triggered the extinctions. Continental seas simply dried up, leaving behind evaporates in huge quantities. We know that great masses of salt were deposited in Permian times in the American West and parts of Europe. Some geologists suggest that the deposition of salt on the continents left the ocean less salty, causing a crisis for marine organisms.

But neither theory—shelf overcrowding or less-saline oceans—accounts for the crisis that affected land life, or for the apparent abruptness of the extinctions. Lately, a theory ascribing the extinctions to extraterrestrial causes has been gaining popularity. The late Paleozoic crisis, like others that have affected life's history, might have been caused by a large meteorite or comet swarm hitting the Earth, a possibility we will discuss in more detail later.

With the end of the Paleozoic, a new chapter in the Northeast's history was about to begin. Two hundred million years of squeeze, squeeze, squeeze, 200 million years of mountain making were over. Now begins a time of stretch—and the making of an ocean.

The Alleghenian orogeny, the Paleozoic's third and final mountain-building event affecting the Northeast, occurred about 290 million years ago during the Permian period. It was caused by the collision of Gondwanaland (including Africa and South America) and the Old Red Sandstone continent. Gymnosperms, which include conifers, began to replace spore-bearing plants as the dominant flora on Earth. Reptiles, which evolved from amphibians, also experienced rapid diversification. The Paleozoic era ended (250 million years ago) with a great extinction of nearly half of all plant and animal species.

Chapter Seven

Birth of an Ocean

Precambrian | Paleozoic | Mesozoic | Cenozoic

4600 545 445 355 250 145 65 2

Millions of Years Before Present

Scale not proportional

Let's take a global look at the arrangement of continents as the Mesozoic era of geological history began 250 million years ago. Rather, we should say "continent"—singular—for 250 million years ago the Earth's surface consisted of one continent and one ocean. We call the continent Pangaea, for "all-earth," and the rocks of the Northeast were then buried at its heart, far from any ocean shore. Geologists call the all-embracing ocean Panthalassa, from the Greek for "all-seas." An arm of the Panthalassa, called the Tethys Sea (Tethys was the mother of Oceanus in Greek mythology), intruded deeply into Pangaea from the east. The Panthalassa Ocean was the predecessor of the present Pacific Ocean, now only a fraction of its former size. Great chunks of the present western United States—the exotic terranes—were islands or plateaus in that former, larger sea; they were gathered up and plastered in layers onto the continent as North America drifted northward and westward.

Pangaea reached from pole to pole. You might, if you had wished, have made a land journey from the North Pole to the South Pole. That journey would have taken you south and west across Siberia into Europe, and then into North America. Turning due south, you would have crossed into Africa near the present site of Boston, left Africa near Nigeria to traverse the bulge of Brazil, returned to Africa, stepped onto Antarctic soil near Durban, South Africa, and walked across an unfrozen Antarctica to the South Pole.

For tens of millions of years, Pangaea had been rotating counterclockwise and drifting northward across the face of the globe. The details of that motion are not clear. It may have involved wholesale remaking of the basaltic ocean floor, with subduction at some places and sea-floor spreading at others. Clearly though, Pangaea moved more or less as a unit, and our region shared that motion. At the beginning of the Triassic Period, 250 million years ago, the Northeast crossed the equator, and by the end of the Triassic, 205 million years ago, had drifted to a latitude of fifteen degrees north, about the latitude at which Central America lies today. Sedimentary deposits from those times indicate a warm climate for our region, with alternating dry seasons and wet seasons, and temperatures uniformly around 70 to 80 degrees Fahrenheit. In many ways, the climate was like that of equatorial East Africa or parts of the southwestern United States today, and (as we shall soon see) the parallels with those regions go even deeper.

Pangaea Disassembled

The similarities between the Northeast in the Triassic and present-day East Africa and southwestern United States go very deep indeed—deep into Earth's mantle, where churning convective motions are today slowly modifying the crust. By late Triassic and early Jurassic times those mantle currents had begun to tear Pangaea apart. The supercontinent that had endured 100 million years began to be disassembled. In the south, Antarctica, India, and Australia broke away from Africa. In the north, South America and Africa, as a block, began to pull away from North America. First, the northern coast of South America unzipped from North America. Then the zipper turned northward and began to separate Africa from land that is now the eastern United States. Africa pivoted on a hinge near Gibraltar, and as it turned it began to close up the Tethys Sea even as the Atlantic opened.

Under the supercontinent of Pangaea hot mantle rock was rising, and where it encountered the cool, brittle crust it was deflected horizontally. Friction between the crust and the moving mantle began to pull the crust apart. The first sign of the coming rupture was a lifting and thinning of the crust. Then came volcanos, hundreds of them, along the line of rupture, where the building pressure in the mantle sought release. The stretched crust fractured along a web of cracks, and through some of the cracks highly liquid lava flooded onto the crust, in some places covering huge tracts of land with layers of glowing rock hundreds of feet thick.

That is how the Atlantic Ocean was born. If you want to find a place in the world today that looks like the Northeast in the Triassic, then go to the rift valleys in East Africa, where a continent is presently being torn asunder. Along the East African rifts, the land has been lifted from below. Linear blocks of crust have downfaulted, broken away, and sagged along north-south-trending faults—creating vertical escarpments and, in some places, several parallel lines of cliffs. The downfaulted sediment-filled rift valleys of East Africa have warm lakes and bubbling hot springs. Not far below the surface, massive reservoirs of seething liquid rock wait to break free.

Near the rift valleys of East Africa stand some of the world's great volcanos, including Mount Kenya and Mount Kilimanjaro. In late Triassic and Jurassic times, the Northeast had volcanic peaks equal in stature. The Ossipee Mountains, just north of Lake Winnipesaukee in New Hampshire, form a circular ridge, a remnant of one of the many imposing volcanic cones that formed as the Atlantic Ocean opened. Other igneous-rock masses near Lake Winnipesaukee and the White Mountains also date from that time.

Rift Valleys

Most of the evidence for rifting in the Northeast has vanished. The decisive rupture where the continents finally broke apart lay 100 miles east of the present coastline, where the continental slope is today. The faults and steep escarpments that paralleled each side of that decisive rift have subsided into the ocean and are now deeply buried under huge thicknesses of sediments on the continental shelf. But two Triassic rift valleys remain very much in evidence, and offer us a tantalizing glimpse of the Northeast as a land of fire beside a newborn sea.

The Newark Basin, cutting across north-central New Jersey from Pennsylvania to the Palisades, and the Connecticut Valley lowlands, between New Haven and the

Massachusetts–New Hampshire line, are floored with sediments and volcanics that were deposited in downfaulted valleys at the time of rifting (Figure 7.1). Reconstructing the ancient landscape is done by inference. The spectacular late Triassic and early Jurassic topography is gone; the precipitous escarpments have been eroded flat, and the gashlike valleys are filled with sediments.

Any visitor to the Newark Basin or the Connecticut Valley will soon recognize that the distinguishing characteristic of many of the rocks is their color. They are reddened by iron oxides that bind the quartz and feldspar grains in the sediments. The oxides indicate a dry, desertlike climate at the time of their deposition. The sediments were washed into the rift valleys from the adjacent uplands during brief wet seasons when rain-swollen streams tumbled onto the valley floors. The rift valleys in the Triassic Northeast, like the rift valleys in East Africa or the American West today (as Death Valley), had no outlets to the sea; in the wet season temporary lakes expanded on the valley floors, and streams carrying sand and gravels down the steep-walled escarpments dumped their burdens in broad alluvial fans. In some places, layers of sediments more than three miles thick accumulated in the downfaulted valleys.

As the escarpments rose and the valley floors slipped downward, the Northeast shook. Sworms of shallow earthquakes in the brittle crust accompanied the wrenching open of the Atlantic. Finally, in the early Jurassic, about 200 million years ago, came floods of lava. Three times in the Newark Basin the Earth opened and layer after layer of the fiery substance from the planet's interior gushed forth. Lava welled up through cracks in the stretched crust and spread out over the valley floors in red-hot liquid sheets hundreds of feet thick. After the first eruptions came a time of relative quiet. The lava layers were buried by sand and mud. A second series of eruptions again flooded the land with lava, and those volcanic rocks too were buried. Then a third. A fourth

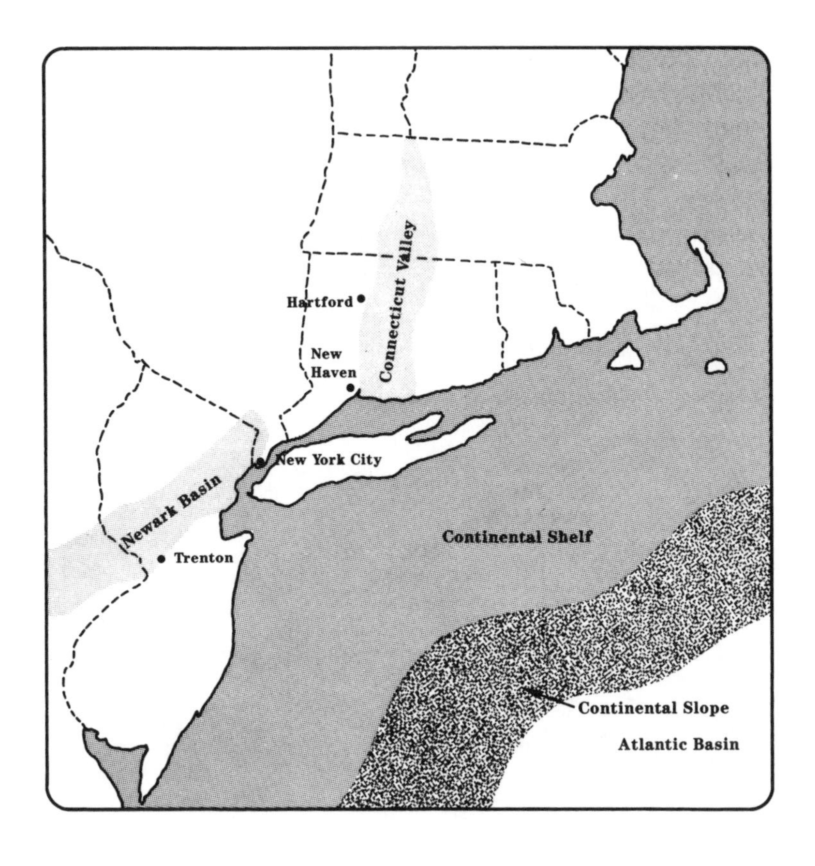

FIGURE 7.1—Rift basins formed in the Northeast during the Triassic. Other rift valleys lay buried on the continental shelf. The final opening of the Atlantic Ocean occurred where the continental slope is today.

flood of molten rock did not reach the surface, but intruded itself between layers of sedimentary rock deep beneath the others, lifting the entire surface of the rift valley by a thousand feet. When these volcanic episodes were over, the valley was filled like a baker's confection, with sediments and basalts in alternate layers. The sediments and the volcanics were deposited or emplanted horizontally.

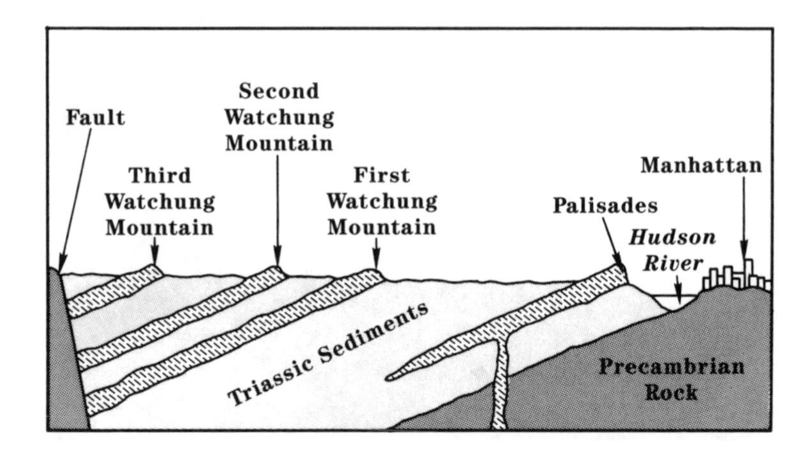

FIGURE 7.2—This cross-sectional view indicates the rocks underlying Manhattan and New Jersey. The hatched formations are lava flows that have resisted erosion.

Today, the entire Newark Basin has been tipped so that the layers slope to the west (Figure 7.2). Erosion has cut across the layers, exposing them at the surface. Where the lava layers are now exposed they have more successfully resisted erosion than have the sediments. The lavas that once flooded the surface of an early Jurassic rift valley now form the three parallel ridges of New Jersey's Watchung Mountains. The deepest and thickest igneous layer, the one that was emplanted beneath all the others, now outcrops on the west bank of the Hudson River to form the majestic Palisades (so named because columnar jointing in the igneous rocks makes the cliff face resemble the vertical logs of a palisade fort). If you stand on the Palisades and look west or south across the New Jersey lowlands toward Pennsylvania, wherever you see a hill or ridge rising above the plain, you are looking at a resistant volcanic remnant from that time of fire.

Similar volcanic episodes disrupted the Connecticut Valley. Here, too, layers of sediment alternated with lava flows, and here, too, the layers were subsequently faulted, tilted, and exposed at the surface. The lava flows are now expressed by ridges that resist erosion. The Holyoke and Mount Tom ranges in the Connecticut Valley are remnants of the lava flows, as are the Pocumtuck Hills. Just north of New Haven, the Wilbur Cross Parkway tunnels through the lava ridge called West Rock. West Rock is steep on both sides and may have been a vertical feeder to now-vanished horizontal layers of lava. When you drive through the tunnel in West Rock, you are penetrating a lava-filled crack in the Earth's crust that formed as the Atlantic Ocean was born.

Dinosaur Tracks

In the Northeast rift valleys, in New Jersey and Connecticut, it is now the resistant volcanic rocks that stand highest and dominate the landscape. But the sedimentary strata interbedded with the lavas hold the key to life in the Northeast at the time when the Atlantic opened.

When glowing lava sheets spread across the rift-valley floors, incinerating vegetation, sizzling into lakes and marshes, any animals that did not or could not run before them were consumed by red-hot rivers of rock. It is highly improbable that meaningful fossil remains could be extracted from volcanic rock. But the sandstones and mudstones that are interleaved with the frozen-lava layers store a rich treasury of Triassic and early Jurassic fossils. Sedimentary-rock exposures in the New Jersey and Connecticut Valley lowlands also bear thousands of footprints made by animals now extinct. The fossils and footprints vividly depict life in the Northeast as the Age of the Dinosaurs began.

The sands and muds of the rift valleys hold many clues to the environmental conditions that prevailed in the Triassic and early Jurassic. Sun-baked mudstones show ripple marks where waves washed lightly across lake shores. Raindrop pits still speckle sandstones. Coal-like remnants of plants that grew in the lakes and marshes give black, sandy shales their color. Thin coal seams in the Connecticut Valley rocks suggest that the margins around the Triassic lakes supported peaty bogs. But among all these "snapshots" of the ancient environment frozen in the rocks, the most dramatic are the footprints.

Everyone has seen skeletons of extinct animals in museums, and you might think that fossil bones are the most prevalent evidence for vanished forms of life. They are not. The bones of extinct animals are much rarer than fossilized animal tracks. In fact, many species are known only from their tracks. Of the more than 150 animal species that have been identified from tracks in the Connecticut Valley mudstones, bones have been unearthed for only about 10 percent. In some ways, the tracks tell us more about these ancient creatures' lives than do their bones. Tracks are made by living animals as they hunt, socialize, or move from place to place. Tracks tell us about posture and pace and behavior.

The earliest documented discovery of fossil footprints anywhere in the world was made in the Connecticut River

Valley near South Hadley, Massachusetts, in red sandstones of late Triassic or early Jurassic age. They were discovered in 1802 by Pliny Moody, a farm boy. (You may recall that Pliny the Elder was one of the great naturalists of classical times.) The three-toed tracks were those of small dinosaurs, but in 1802 nothing was known of that ancient race of animals. Naturally people assumed the tracks had been made by giant birds, possibly even (someone said) Noah's raven. We know now that birds had not yet evolved at the time the tracks were made, although small dinosaurs were beginning to experiment with flight.

The track-making dinosaurs of the Connecticut Valley were not the familiar giants that would dominate the late Jurassic and Cretaceous periods, such as the *Brontosaurus*, *Tyrannosaurus rex*, or *Triceratops*. Most of the Triassic and early Jurassic dinosaurs were smaller than a human being and walked erect. From the tracks we can see that many of them traveled in packs. At a fossil trackway near Holyoke, Massachusetts, prints made by at least nineteen individuals of the carnivorous genus *Eubrontes* have been identified, all moving together from east to west in the general direction of Mount Tom (Figure 7.3). *Eubrontes*, among the largest of the Connecticut Valley dinosaurs, was perhaps twenty feet long, and some of the tracks made on an ancient lake's muddy bottom suggest that the creature could swim.

Dinosaur tracks can be readily viewed at several locations. One site, owned by the Massachusetts Trustees of the Public Reservations, is just off Route 5 near Smith's Ferry. A broad exposed ledge bears tracks of about a dozen dinosaur species. The best place to study fossil tracks and the creatures that made them is at Dinosaur State Park in Rocky Hill, Connecticut. This site was discovered in 1966 during excavations for a Highway Department laboratory. More than a thousand dinosaur footprints are on view at Rocky Hill, along with exhibits showing how the tracks were made.

How do dinosaur footprints end up in solid rock? Two

FIGURE 7.3—Triassic dinosaurs make tracks in the mud of the Connecticut Valley.

hundred million years ago, at the present site of Rocky Hill, dinosaurs left their prints on the muddy bottom of a shallow pond. The pond dried up and the footprints were baked hard by the sun. A new period of wet weather washed sediments of contrasting texture onto the pond floor. These sediments also hardened. Sedimentation continued until the tracks were deeply buried and the original muds were turned to stone. Later, when the entire region was uplifted and eroded, the sedimentary rocks broke apart along the original bedding planes, exposing the tracks.

Many dinosaurs that left their footprints in the Connecticut Valley have also been identified in the American Southwest and in Europe. These creatures seemed to have ranged widely, and until the rifting that opened up the Atlantic (about 200 million years ago) they found no impediment to passage to and from North America, Europe, and Africa. *Plateosaurus*, for one, has been found on most of the continents—it was one of the cosmopolitan travelers of Pangaea.

Sedimentary rocks in New Jersey have also yielded relics from the time of the dinosaurs. One fossil find is of special interest. In 1960, Alfred Siefker, an amateur geologist, and two of his high school friends were exploring the Granton Quarry in North Bergen, New Jersey. They found a fossil that they immediately recognized as different from anything else they had seen. The little creature, now called *Icarosaurus siefkeri*, had greatly elongated ribs. It is generally agreed that the ribs supported membranes of skin that enabled this little lizard to glide through the air. *Icarosaurus siefkeri* is the earliest known vertebrate that could achieve any kind of flight.

The Granton Quarry is a particularly fruitful site for uncovering fossils that help us understand life in the late Triassic rift valleys. Many fish have been found there, including *Diplurus newarki*, an extinct member of the coelacanth group of fishes. These primitive lobe-finned fish were long thought to be extinct, until one was unexpectedly pulled up from deep waters near Madagascar in 1952. The New Jersey coelacanths undoubtedly lived a precarious life in ponds and marshes that were forever close to drying up. The Granton Quarry also yielded a fossil of the crocodilelike phytosaur *Rutiodon* (Figure 7.4). Phytosaurs seem to have been common in the late Triassic marshes. With their long, thin jaws lined with rows of sawlike teeth, they were a formidable threat to any fishes or small reptiles that came their way. Today, almost the entire area from which the Granton Quarry fossils were unearthed is buried under shopping malls and industries.

On a New Shore

Throughout the late Triassic and early Jurassic, Earth's brittle crust was stretched. A last decisive rupture opened up the Atlantic, and new sea-floor crust began to form along a midocean rift. The decisive break might conceivably have come along the rift valleys in New Jersey and the Connecticut Valley. If it had, then all of coastal New Jersey and New England would today be a part of Africa. But it did not happen: The decisive split occurred several hundred miles to the east, where today the continental shelf falls precipitously to the ocean floor. There the Atlantic Ocean was born.

During these events, the Northeast would have been a violent place in which to live. Catastrophic lava flows of the early Jurassic must have made a frightening image. The earth opened up and poured huge quantities of liquid rock out upon the land. Nothing comparable has occurred anywhere on earth in historic times. During these prodigious eruptions, the rift valleys were shrouded by thick clouds of poisonous gases, expelled with the lava. The earth shook with quakes as the molten rock ascended through pipes and fissures to the surface. The lava buried the landscape—marshes, mud flats, lakes, savannas—beneath tens or hun-

FIGURE 7.4—The crocodilelike phytosaur *Rutiodon* was an inhabitant of New Jersey Triassic marshes.

dreds of feet of luminous, flowing rock. The more agile dinosaurs ran before the flood, seeking escape in the uplands to the east and west of the rift valleys. *Phytosaurs* and fish were sizzled where they were overtaken. Slowly the thick sheets of lava cooled and became solid rock. Even more slowly did surface weathering, subsidence, and sedimentation re-create in the valleys conditions allowing repopulation by plants—mostly conifers and ferns—and animals. Then, after millions of years of relative quiet, came another lava flood, and then another.

Earth's crust ripped open from south to north. At first the scene resembled East Africa today, or southern Nevada. Then, as the Atlantic widened, the geography of the Northeast came to resemble that of the Red Sea or the Gulf of California, places where today continents are being stretched and broken, and new oceans are being born. As the Atlantic widened at the midocean ridge, and as new basaltic crust was extruded inch by inch from within Earth's fiery furnace, salt water flowed from the south to fill the new basin. The sea grew to become an ocean, and the margins along the newly defined continents subsided. Faulting, earthquakes, and lava floods ceased. The more easterly of the North American rift valleys are now buried on the continental shelf. The valleys farthest inland, the Newark Basin and the Connecticut Valley, were filled to the brim with the last eroded remnants from the nearby highlands. North American dinosaurs were separated from their African cousins, and from that time forward their evolution would take place in isolation.

Again, Punctuation from Space

The story of how the Atlantic Ocean opened spans two great geologic periods: the Triassic and the Jurassic. In rock types and the tectonic events they record, the Northeast shows no clear boundary between the two periods. But here, as elsewhere, a sharp discontinuity appears in the fossil record, and it is the fossil discontinuity that defines the boundary between the two periods in geologic time. At the Triassic-Jurassic boundary came yet another mass extinction. Fully 40 percent of the animal families whose fossils are found in late Triassic continental rocks are missing from the Jurassic sediments just above them. Twenty percent of the families of creatures that lived in the sea also quite suddenly vanished. Clearly, something happened 205 million years ago that had disastrous consequences for life on Earth.

In summer 1985, Paul Olsen and Neil Shubin, scientists from Columbia and Harvard universities, unearthed a fossil trove in rocks exposed by low tide on the floor of Nova Scotia's Bay of Fundy. The rocks were mudstones and sandstones, interleaved with volcanic basalts—rocks typical of the New England rift valleys of the late Triassic and early Jurassic. The Jurassic rocks contain the fossil bones of tritheledonts, a group of reptiles closely related to mammals and previously known only from Africa. Also plentiful are sphenodonts, small lizardlike reptiles whose only living relative is the tuatara of New Zealand, and yard-long "saber-toothed" crocodiles with spindly legs and a whiplike tail. Penny-sized footprints of a dinosaur no larger than a sparrow are also found in the rocks. But absent in the Jurassic formations are almost half the reptilian and amphibian creatures known from the Triassic rocks just below. The Bay of Fundy fossil hoard is a vivid archive for the catastrophic event that terminated the Triassic.

Many geologists now believe that the Triassic-Jurassic extinctions may have been caused by a massive meteorite that collided with Earth (we have more to say about meteorites at the end of chapter 8). A possible scar from that collision is the monstrous Manicouagan impact crater in central Canada, 500 miles northwest of Nova Scotia. The sixty-mile-wide crater appears to be old enough to qualify as the Triassic-Jurassic impact site. The dust thrown up by a meteorite collision of this magnitude would have thrown the rift valleys of the Northeast—indeed the entire planet—into temporary cold and darkness.

> *As the Mesozoic era began (250 million years ago), the world's land masses had joined and formed Pangaea, the supercontinent. During the Triassic and Jurassic periods (250–145 million years ago), Pangaea began to break apart and the Atlantic Ocean was formed along a new mid-ocean rift. Extensive volcanic and sedimentary deposits from these periods exist in the Newark Basin and the Connecticut Valley. Within their rocks are the fossils of many dinosaurs that inhabited the Northeast, including* Icarosaurus siefkeri, *the earliest known vertebrate capable of flight.*

Chapter Eight

Quiet Before the Storm

Precambrian | **Paleozoic** | **Mesozoic** | **Cenozoic**

4600 545 445 355 250 145 65 2

Millions of Years Before Present

Scale not proportional

The era of fire subsided. The great lava sheets cooled. The volcanic gas that had wrapped the Northeast in an acrid cloud dissipated. The waters of a bright new ocean lapped at a new shore. Hundreds of miles east, rifting continued, underwater, along a fiery gash on the ocean floor. Out there, hidden from view by water miles deep, the Atlantic continued to widen, even as it does today. But in eastern North America, the time for tectonic upheaval had passed.

When the North Atlantic Basin opened, it unzipped from south to north. First, rifting separated North America from South America, where the Caribbean is today. Then the zipper began to cut between South America and Africa, as one arm in a three-way split called an aulocogen, but failed. The split's successful northern arm pulled North America away from Africa. Occasionally waters from the Pacific and the Tethys Sea spilled into the new rift valleys and evaporated in the hot climate, forming thick salt sequences that now lie buried by sediments on the continental margins. With continued rifting, a permanent connection between the Pacific Ocean and the Tethys Sea was established near Gibraltar, breaking Pangaea into two parts: in the south, the continent Gondwana, embracing South America, Africa, Australia, India, and Antarctica; and in the north, Laurasia, including North America, Europe, and Asia. But the rifting was not finished. During Cretaceous times (145–65 million years ago) convection in the mantle would finish tearing Pangaea into pieces.

Rifting in the North Atlantic pushed hard to bisect Laurasia. A gap opened between Spain and France that would become the Bay of Biscay, but rifting there was inconclusive—another failed arm in a three-way split. The active arm of that rift system sliced north like a knife, and cut Ireland, Britain, and Scandinavia away from Labrador. Then another three-way split developed. One arm, cutting northwest, almost pared Greenland from Canada. Then a more decisive rupture detached Greenland from Europe, connecting the Atlantic Ocean with the Arctic Ocean. North America was on its own.

During the final stages of the disassembly of Pangaea, after 300 million years of relatively steady tectonic violence, the Northeast had a time of quiet. As the North Atlantic widened, the region drifted away from any active plate boundary. Dramatic mountain building and ocean making gave way to uninterrupted erosion, the gentle work of wind and water. The Appalachian Highlands, rejuvenated during the Triassic, came tumbling down again, grain by grain. Every crystal of frost was a wedge that pried another rock fragment from a mountain peak. Every raindrop dissolved a bit of rock and carried it to the sea. Erosion is usually slow, unspectacular work. The story of the Northeast in the late Jurassic and Cretaceous has a quiescent quality about it, like a leaf fluttering to the ground. But, as we shall see, the story ends with a bang.

Bringing the Mountains Down

Given enough time, erosion will level any mountain, flatten every hill, erase every bump and ripple in the land. If it were not for periodic crumplings or liftings of the continents, the land would everywhere be as flat as the sea, at the level of the sea. The average rate at which erosion works over North America is about two inches every thousand years. Mount Washington, in New Hampshire, is presently our region's highest mountain; it rises 6,288 feet above sea level. If erosion cuts Mount Washington down at the continental average, in 30 million years the mountain will be gone. If uplift ceased, in 30 million years the Northeast would be as flat as Kansas.

The rate of erosion depends on climate and the composition of the rocks being weathered. Consider the Egyptian obelisk that stands in New York's Central Park. That hieroglyphic-inscribed granite monolith was brought to the United States in 1879. When the obelisk was placed on its

New World pedestal, the inscriptions were nearly as sharp and crisp as if they had been chiseled that very day. The hieroglyphics had survived unscathed in the dry Egyptian desert for nearly 4,000 years. But cold winters—mixed with many freezing and thawing cycles hot, humid summers and damaged the monument within a few years. Many inscriptions on the lower part of the obelisk are now practically unreadable. If erosion can remove millimeters from the obelisk's surface in only decades, ten thousand years would see the obelisk gone.

During late Jurassic and Cretaceous times, the climate in the Northeast was warmer and more humid than it is today, both because the region was 15 degrees closer to the equator, and because climate over the entire planet was milder. The generally warmer climate can be partly attributed to the higher sea levels that resulted from widespread oceanic rifting; water is less reflective than land and more efficiently absorbs energy from the sun. Also, more carbon dioxide in the atmosphere, produced by increased global volcanic activity, acted like glass in a greenhouse to retain the energy of sunlight. But if frost was absent as an eroding agent, chemical weathering was not. Water in the atmosphere combines with carbon dioxide to form dilute carbonic acid. Acid rain is an effective weathering agent, especially in a warm environment where chemical reactions are speeded up. The rains fell on the Northeast highlands, and the rock dissolved and fell apart. Erosion was unrelenting. Within 50 million years, the mountains that had been lifted in Triassic times were mostly leveled.

By the late Cretaceous (100–65 million years) a broad, flat erosional surface, called a *peneplain*, stretched across central and southern New England. Here and there the monotonously flat landscape was broken by a resistant remnant from the old uplands, composed of rocks with mineral characteristics that made them especially strong and chemically inert. One of those remnants, New Hampshire's Mount

Monadnock, gives its name to mountains that have resisted degradation to a plain; geologists now refer to any mountain that stands atypically above an erosional surface as a *monadnock*. The obdurate summits of Mount Acutney, Mount Wachusett, and Mount Greylock also forestalled diminishment. Like Mount Monadnock, they remain today the dominant features in their landscapes.

The sand, grit, and silt grains that were eroded from the surface of the Northeast during the Cretaceous were carried by streams and meandering rivers to the sea, where they were deposited in a sloping wedge of sediments on the continental margin. On the outer continental shelf the deposits were more than two miles thick. Carbonate reefs and lagoons on the Jurassic continent's margin were gradually silted up and buried. Today, the Cretaceous sediments are infrequently exposed, and so we have only rare glimpses of any fossils that might be within them. On the continental shelf, under the present Atlantic, the sediments are buried by more recent debris that has been carried into the sea from the continent. But one band of Cretaceous sands and muds remains unburied, and the New Jersey Turnpike runs down its middle, from Raritan Bay to the Delaware River. Perhaps the finest exposure of Cretaceous sediments occurs where this band is cut by the sea, on the south shore of Raritan Bay, between Cliffwood Beach and Highlands, New Jersey. It is to the New Jersey sediments that we must look for the fossil record of life in the Cretaceous.

The sediments along the Jersey Pike mark approximately where land met sea during Cretaceous times. From here, the ancient shoreline curved across Raritan Bay and followed the line of Long Island, Block Island, and Martha's Vineyard. From the mouth of New York Harbor eastward, the Cretaceous deposits are now covered by water, or by sands and gravels piled up by glaciers during the recent ice ages. But here and there, evidence of Cretaceous New England pokes above the younger detritus, providing an occasional tantalizing peek at an otherwise lost landscape. Outcrops of Cretaceous deposits lie near Plainview, Long Island, and on Long Island Sound near Oyster Bay. One of the most spectacular outcrops of Cretaceous clays (and even a bit of low-grade coal) occurs in the fabulously colored cliffs at Gay Head on Martha's Vineyard.

In Flowering Swamps

The abundant fossils in exposures of Cretaceous sediments in New Jersey allow us to reconstruct the Northeast coastal habitat of older times. It was a wet world, with the sea advancing and retreating upon a gently sloping shore. Rivers flowing from the New England uplands meandered across a broad plain, and where land met sea were extensive swamps and marshes.

Plant life on the plain and in the marshes was dominated by conifers and ferns. Even today, Cretaceous pine cones tumble from the cliffs at Cliffwood Beach, New Jersey. Conifer pollen and fern spores are common in New Jersey clays and have been found in bore holes on Long Island. A tangled forest of those plants found its footing on the wet shore. And in the Cretaceous clays and sands is evidence of a remarkable transformation of Earth's face—fossilized flowering plants. Flowers! Magnolias bloom among the pines and sassafras among the ferns, gaudy new additions to a brown and green world. Sweet new scents permeate the warm and humid air. The upland hills are decked with color. Deciduous trees appear for the first time. And something else: In the youngest of the Cretaceous coastal sediments are fossilized seeds of grass! The grasses did not prosper until the following geologic era, but in the Cretaceous they made their debut. Meadows. Green plains. It was a time of softening.

In the shallow waters on the continental shelf and in the muddy estuaries and bayous flourished a rich fauna. Oysters and ammonoids. Schools of fish. Giant sea turtles. Many-armed invertebrates, now extinct, related to the squid and octopus. Sharks with teeth the size of your hand. And most dramatic of all, and most unlike anything living today, the plesiosaurs and mesosaurs (Figure 8.1). These "sea monsters" patrolled the shallow waters, preying on almost anything that moved. The deadly streamlined plesiosaurs, up to thirty-five feet long, propelled themselves through the water with two strong pairs of paddlelike limbs. Their bones have been found at Woodbridge, New Jersey. Mesosaurs were probably the most fearsome predators in Cretaceous seas. These huge marine lizards grew to forty or fifty feet long. They were powerful swimmers and their jaws were lined with razor-sharp teeth.

The fossil record of animal life on land is sketchier. Wasps and ants inhabited New Jersey by late Cretaceous times; the ants were the earliest social insects. The clays at Cliffwood Beach have yielded a bit of amber, fossilized tree resin, encasing like Snow White's transparent coffin the remains of several insects, including two worker ants. The first fairly complete dinosaur fossil skeleton found in the United States was unearthed in Cretaceous clay eight miles east of Philadelphia; the reconstructed skeleton of that thirty-foot-long duck-billed hadrosaurus can now be found in the New Jersey State Museum in Trenton. Which other dinosaurs inhabited the northeastern coast of North America, or the Appalachian Uplands, may never be known. The sediments that held their bones have been eroded away and are deeply buried on the continental slope. Almost nothing is left from those times except a thin clay strip down the middle of New Jersey. The Cretaceous fossil record of terrestrial life in the Northeast is slim pickings.

Did *Triceratops* move in herds on the New England plains? Did *Tyrannosaurus rex* snatch up New Jersey's first mammals in its terrible claws? Did leather-skinned pterosaurs take wing from cliffs in the ancestral White Mountains, struggling to match the agile flight of the feathered *Archaeopteryx,* one of the earliest birds? The sedimentary deposits

FIGURE 8.1—''Sea monsters'' of Cretaceous times: above, two plesiosaurs; below, a mesosaur

in the American West have yielded a rich record of Cretaceous animal life, but in the Northeast it was a time of erosion rather than deposition. The eons have all but obliterated the Cretaceous surface from our region, stripping it from the land and dumping it at sea. Only that little ribbon of ancient shoreland in central New Jersey, or the occasional outcrops on Long Island or Martha's Vineyard offer hope of more extensively reconstructing Cretaceous life on our northeastern shore.

Rising Seas

Throughout the late Jurassic and Cretaceous, sea levels rose around the world. At the end of the Cretaceous period the sea level with respect to the land was as high as at any time since the Precambrian. Broad areas of the continents were submerged. In North America, a shallow sea, called the Sundance Sea, covered most of the continent's center, extending from the Arctic Ocean to the Gulf of Mexico, down across the region of the present Rockies; if you had lived in North America in those times, you would have had to travel by boat from Missouri to Nevada. The reasons for these deep trangressions of the sea onto the continents are complex. North of our region, Europe was being torn away from North America. To the south, Africa was unzipping from South America. India, Antarctica, and Australia were being detached from one another and from Africa. Pangaea was being shredded by that huge tectonic engine, the Earth's churning mantle. Five hundred miles to the east of our northeastern shore, the Atlantic Ocean floor was being lifted from below and wrenched asunder. Rapid rifting and seafloor spreading all over the Earth raised lofty ridges along the spreading axes. Build mountains on the sea floor and the volume of the ocean basins must decrease. And where does the water then go? Up! And it overlaps the land.

Toward the end of the Cretaceous, rising seas had pushed far inland, across the flattened Northeast peneplain.

Perhaps the summit of Mount Monadnock stood as an island in the sea. How far inland the sea extended is not known. Sediments were surely deposited on that shallow sea floor, but those muddy beds have long since been erased. In Europe, a thousand miles across the Atlantic, the continent was also flooded. The waters there were astonishingly rich in planktonic life. The calcite skeletons of microscopic organisms settled in vast numbers onto the sea floor and formed the thick chalk beds that are today exposed at the White Cliffs of Dover and elsewhere along the English Channel. In the waters off the New England and New Jersey coasts, similar microorganisms undoubtedly flourished, but apparently in lesser numbers, and any capacity they had to make (in their tiny dyings) beds of chalk was diluted by the great flood of sand and mud that came pouring off the continent.

In a sense, a continent's erosion is never completed. Recall that Earth's crust floats on the "mushy" upper-mantle rock. When a layer of rock is removed from a continent by erosion, the continent rises—just as a raft rises in the water when you step off—exposing new rock to weathering agents. When detritus from an eroded continent is dumped onto a continental shelf or into a nearby ocean basin, the weighted crust sinks, making room for new sediments. The continent gradually thins and rises, while the margin of the continent thickens and sags from the ever-growing burden of sediments. The work continues until a continental collision occurs, at which time the trough of sediments at the continent's edge will be folded back up into new mountains. Every new mountain range is built from the crumbs of the range that went before. The eroded fragments from yesterday's Appalachians are today on the Atlantic Ocean floor off the North American coast. And hundreds of millions of years from now, these same sediments—and the marine fossils they hold—will be lifted into lofty peaks in a refurbished range.

The Great Extinctions

The Cretaceous in the Northeast was a quiet time. The land and sea rose and fell in gentle cycles. It was a time of slow evolutionary innovation: the flowering plants diversified; the mammals consolidated their tenuous evolutionary niche in the dinosaurs' shadow. But the Cretaceous ended with one of the most dramatic extinctions ever. The Northeast shared this calamity with the entire planet. Of the animals that left a permanent record in the Northeast Cretaceous sediments, almost all became extinct. Like all its dinosaur cousins, the duck-billed hadrosaurus disappeared, as did the great marine reptiles, the plesiosaurs and mesosaurs. Most land animals with a body weight of more than fifty pounds were exterminated. In the seas, huge numbers of species, at every level of the food chain, vanished. The only clue to the cause behind this catastrophic event is a thin layer of clay, rich in soot and the rare element iridium, which is found worldwide in the sedimentary record. From this layer, scientists have proposed a widely accepted hypothesis to explain the mass extinctions.

Always, the Earth has been bombarded by meteorites from space. Meteoric scars can be found in some of the planet's oldest exposed rock. It appears that every million years or so, on the average, the planet is hit by an object capable of excavating a crater five miles in diameter. Every ten million years something still larger will strike. Sixty-five million years ago, according to the asteroid-impact hypothesis, an exceptionally big chunk of rock came hurtling out of space to collide with the Earth (Figure 8.2).

This hypothetical meteorite, three to ten miles in diameter, was traveling at tens of thousands of miles per hour. It carried energy equivalent to that in 100 million hydrogen bombs. The meteorite passed through Earth's atmosphere and ocean like a bullet through tissue paper, excavating a hole in the crust the size of Rhode Island. Because the crater

FIGURE 8.2—Scientists speculate that the collision of a large meteorite with Earth caused the final extinction of the dinosaurs. The flying reptiles are pterosaurs. Condylarth, below, was an early mammal.

has not been identified, it must be assumed that the meteorite struck the sea floor. If so, it raised a tidal wave more than a mile high that rolled across the broad, flat coastal plain and washed against the Green Mountains and White Mountains. Half the meteorite's energy was transferred to the atmosphere, giving rise to a short yet lethal heating pulse. Land animals were particularly vulnerable to this sudden rise in temperature. Firestorms were ignited. Nitric acid produced in the atmosphere by the heating pulse raised the acidity in lakes and seas. A huge mass of material, part meteorite and part Earth crust, was lofted into the atmosphere by the impact. The material in the pulverized meteorite contributed the rare element iridium to the subsequent fallout. Winds soon carried the dust from the impact and the soot from the firestorms around the world, wrapping the entire planet in a dark shroud. Sunlight was blocked. For a year or more the Earth cooled, the oceans by a few degrees, the land areas by more. A long winter enveloped the planet.

For several months light at the Earth's surface was below the minimum for dark-adapted human vision. The land and the waters were plunged into darkness by the dust veil that covered the planet. Photosynthesis was disrupted, and land plants began to die. Animals, especially large ones with greater food requirements, had trouble finding nourishment. Eventually they starved. The temporary cessation of photosynthesis had particularly disastrous consequences in the oceans. The surface-dwelling microplankton, upon which all sea life ultimately depends, could not survive for more than a few weeks without sunlight. Once these planktonic organisms died, the entire oceanic food chain collapsed.

This is only one theoretical version explaining the calamity that ended the Cretaceous period. Perhaps the extinctions were not as sudden as some have supposed. Vigorous volcanic activity, cooling climate, and regressing seas accompanied the end of the Cretaceous and may have been instrumental in the extinctions. Though geologists will continue to debate the nature and cause of the disaster, its magnitude is clear. The dinosaurs' long dominance came to an end. Never again would a member of that race leave footprints in the Northeast sands and muds. But the story is not without a silver lining: Some of the hardier or more resourceful forms of life that had their origin or time of perfection in the Cretaceous—the flowering plants, the grasses, the mammals, the birds, the social insects—survived the disaster and went on to prosper during the Cenozoic era (meaning "recent life").

Among the most important of these groups was the mammals, hairy, warm-blooded animals that bear their young live and nurse them on milk. They made their first appearance during the time of the dinosaurs. The Cretaceous mammals were small, similar to a possum or shrew but no larger than a modern rat. These early mammals had one special advantage that held promise for the future and helped compensate for their pigmy dimensions (compared to the dinosaurian giants): Their brain was relatively larger than any other creature's. It is tempting to imagine these resourceful little beasts skittering between the feet of a thunder-footed tyrannosaurus rex and other carnivorous dinosaurs, struggling to survive among those monstrous and powerful predators. When the mass extinctions came as the Cretaceous ended, the mammals, deep in burrows, temperature-regulated, fur-warmed, big-brained, and clever, somehow managed to survive while their gargantuan tormentors perished. Freed from competition with dinosaurs, the mammals experienced an astonishing proliferation. Within ten million years of the Cretaceous-Tertiary extinctions, most of the modern orders of mammals had come into existence—including the primates, our own ancestors.

The Cretaceous was a time of erosion in the Northeast, and large amounts of sediment were deposited on the margin of a widening Atlantic Ocean. Flowering plants and grasses evolved, as did mammals and birds. The most impressive denizens of the Cretaceous, however, were the dinosaurs, including Tyrannosaurus rex *and* Triceratops. *A thin layer of the rare element iridium, found in sediments 65 million years old, suggests that a large meteorite collided with the Earth as the Cretaceous closed. The resultant dust cloud, though still a theory, may have caused this period's mass extinctions.*

Chapter Nine

Age of Ice

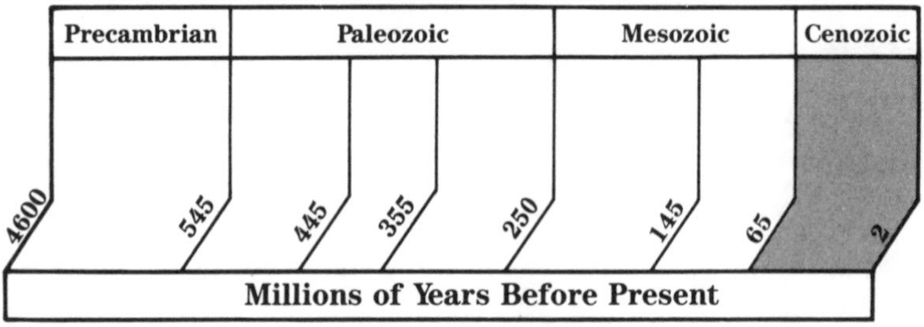

Scale not proportional

If by unquiet we mean those crust-shattering events which accompany the merging and rending of continents, then the Northeast has been tectonically quiet since the Triassic. But continental drift had not ceased when the New Jersey and Connecticut lava floods grew hard and cold, nor has it ceased yet. Throughout the Cenozoic, the Atlantic Ocean has continued to widen, opening like a door hinged at the tip of Greenland, opening a few centimeters a year; the Atlantic Ocean has grown wider by a thousand miles since the extinction of the dinosaurs. At about the same time as the Cretaceous-Tertiary catastrophe, Greenland was wrenched away from Canada, opening the Labrador Sea. That rifting of the crust proved inconclusive, and about 50 million years ago the opening of the Atlantic turned northeastward and slashed between Greenland and Europe. The Iceland rocks oozed up from the midocean ridge; then, as now, that island of volcanos and hot springs might more appropriately be called "Fire-and-Iceland."

Farther afield, the Cenozoic saw the Earth shaping itself on a global scale. Australia ripped away from Antarctica and began its slow northward drift. India moved northward and eased against the underside of Asia, raising the Himalayas, the highest mountains on Earth. Africa squeezed Europe, folding up the Alps and wringing out the last of the Tethys Sea like water from a twisted rag; the famous Matterhorn is a bit of Italy pushed by the squeeze of continents onto the top of Switzerland. In western North America, the growing Rockies lifted ancient sediments (and the bones of swamp-loving dinosaurs) high into mountains.

In the early Cenozoic, the peneplain that had formed in the Northeast during the Cretaceous began arching upward. Rivers were rejuvenated and immediately began etching the peneplain surface with valleys. Eventually the uplift ceased, and by the middle of the Tertiary period (about 40 million years ago) a new featureless erosional surface had formed, called the Schooley Peneplain. This plain, too, was coursed by meandering rivers. On the present continental shelf, and along the New Jersey coastal plain, the old continental margin was ever more deeply buried.

After the Schooley Peneplain formed, the crust of the continental interior again began to rise, and once more the rivers started to entrench themselves. The peneplain surface began to be cut away, back down toward sea level. Where resistant quartzite rocks intersected the plain, ridges endured. Where less resistant carbonate rocks lay, valleys formed. The ridges and valleys of western New Jersey and Pennsylvania let us glimpse the roots of a once vast mountain range, again and again planed away toward sea level, and again and again renewed by the uparching continental crust. Some tens of millions of years from now—barring more lifting—the New Jersey and Pennsylvania mountains will again be smoothed away to a featureless plain.

A similar sequence of events occurred in southern New England. If you travel by automobile in southern New Hampshire or central Massachusetts, you feel you are moving in a rolling, broken landscape. But if you climb one of the mountains that rise above the landscape (one of the monadnocks) and look toward the horizon, you will see that the hilltops are all about the same height and merge to give the appearance of an almost unbroken plain. The hilltops are the last bits from an ancient peneplane surface that is even now being cut away.

In the Northeast, only along the Jersey shore, where on several occasions seas intruded deeply upon the land, have Tertiary sediments collected in an accessible place. Here, in the coastal New Jersey sands and clays, can be found a fossil record of life in Tertiary seas. A plethora of small shelled creatures turn up, including brachiopods, gastropods, and corals. But you also find snakes—twenty feet long—and sea turtles, and crocodiles, which left their bones in the sediments when the rising and falling sea moved back and forth across the swampy continental margin.

Boulders out of Place

The flowing water that dissected the Schooley Peneplain was the chisel and the gouge that gave the present Northeast topography its "rough cut." Moving ice provided the polish. For much of the past three million years the northeast landscape has been buried by glaciers, as even today Greenland and Antarctica are mantled in ice. The middle of the continent-spanning North American ice sheet was in central Canada, near Hudson Bay (Figure 9.1). For an ice cap to grow, more snow must fall in winter than is lost by melting in summer. Then, year by year, the snow cover will deepen and turn to ice. When the ice is thick enough (a mile or so thick), its own weight will cause it to flow outward along the glacier's perimeter. For much of the past three million years, conditions in central Canada have been right for an ice sheet to grow. Again and again the ice cap expanded and contracted; again and again mile-high glaciers extended their frozen grip down across the continent. As the ice advanced, it pushed all before it, like a bulldozer, scraping and breaking and polishing the land. As the ice retreated—melting back faster than it edged forward—it dumped its load of scraped-up erosional debris upon the polished bedrock. Today hardly a place in the Northeast does not show the shaping influence of ice.

On December 21, 1620, the Pilgrims alighted from the Mayflower at Plymouth, Massachusetts, and, according to tradition, made the first landing on a rock that has become enshrined in American folklore. Like the Pilgrims themselves, Plymouth Rock is a traveler. It is a glacial erratic, a boulder plucked up by moving ice far to the north and dropped at the place where the Pilgrims found it. On the geologic time scale, Plymouth Rock, like the Pilgrims, is a recent arrival on the Massachusetts shore.

Erratic boulders—boulders that are dissimilar to the bedrock in the region where they are found—intrigued geologists for a long time before they recognized the reality of the

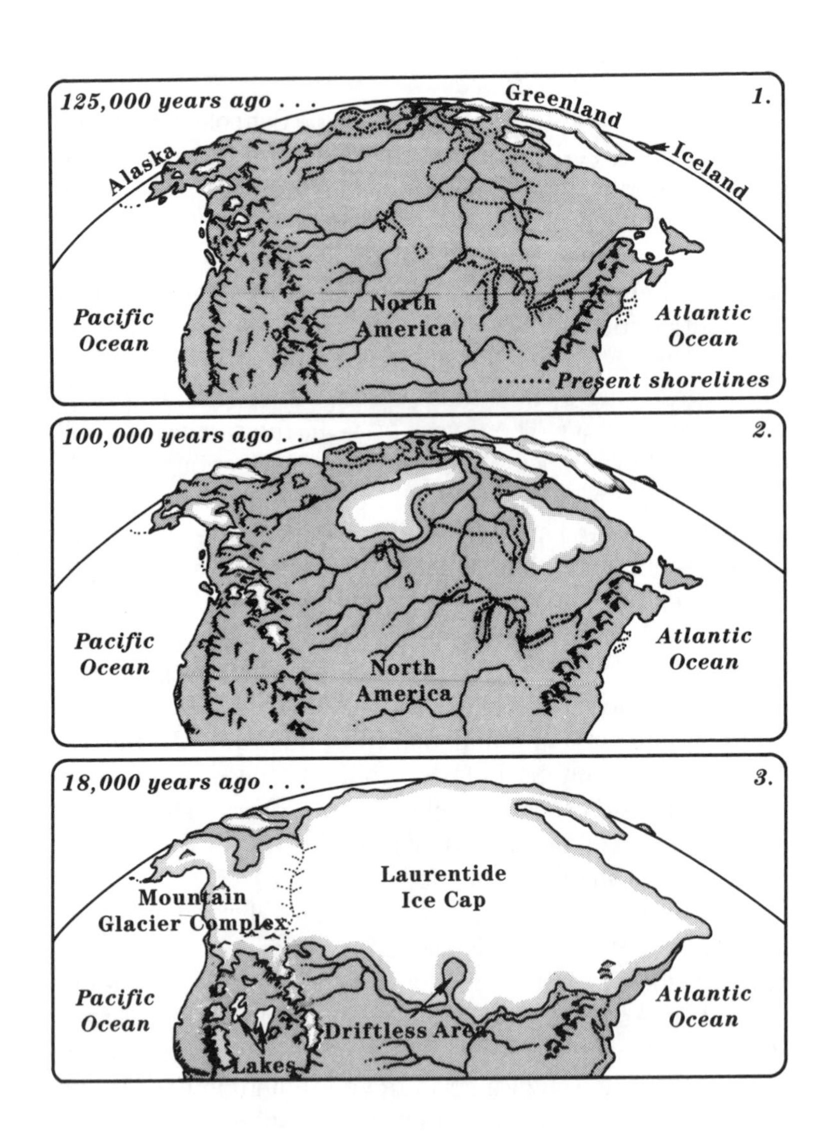

FIGURE 9.1—The advance of the Laurentide ice cap reached its greatest extent about 18,000 years ago. Sea levels were lower because large amounts of water were frozen.

ice ages. Such boulders are common over many parts of the British Isles, northern Europe, and the northern parts of North America. Impressive glacial erratics that are easily accessible to the traveler are Bartlett Boulder and Sawyer Rock on Route 302 near Bartlett Station, New Hampshire, and Monhegan Rock north of Montville, Connecticut. In the early 1800s many thought these boulders had been carried to their present positions by drifting icebergs when the sea covered the land, possibly during the biblical Deluge. According to this view, when rock-bearing icebergs drifted south into warmer waters and melted, they dropped their passenger boulders in odd places. These erratic boulders, and the other out-of-place sediments usually associated with them, came to be known as drift, which fits their presumed iceberg origin and is still used today.

But the iceberg theory failed to answer some puzzles about erratic boulders. The largest glacial erratic boulder in our region, and one of the largest in the world, is in Madison, New Hampshire. It is 83 feet long and weighs 5,000 tons, as much as an ocean freighter. Mineralogical examination of the boulder shows that it originated as a bedrock outcrop two miles north of the place where it is presently found. No conceivable combination of events involving rising seas and drifting icebergs could satisfactorily explain how this gigantic piece of the Earth's crust could be broken off and transported.

Midway through the nineteenth century an alternative explanation for the glacial erratics was posed by Jean de Charpentier, Louis Agassiz, and other astute observers, from the evidence visible in the boulders themselves, various other kinds of "drift," and scratches on bedrock. These scientists proposed that at some time in the not too distant past great parts of the continents were spanned by glaciers. Exploration on the Greenland ice sheet in the 1850s helped convince geologists that vast continent-spanning ice sheets are possible, and that they bring about exactly the kinds of erosional and depositional features that are now found in many unglaciated parts of the northern continents. The Greenland ice sheet clearly has enough power to pluck up and carry the ship-sized Madison Boulder. By the 1860s, most geologists in Europe and North America were convinced that parts of both continents had once been covered with a Greenlandlike sheet of ice.

A Cooling Planet

The Earth's surface is delicately poised between fire and ice. Some astronomers estimate that, if Earth were only 10 percent closer to the sun, a runaway "greenhouse effect" would have turned the planet into the hellish inferno that is Venus. Water vapor in the air, evaporated from the ocean, would have trapped the sun's heat, like the glass in a greenhouse. The heat would have further vaporized the seas, cloaking the planet with even denser clouds of heat-trapping moisture. The planet's surface, including the present sea floors, would have been a scorched, lifeless desert. Water would remain only as steam in the air. On the other hand, if the planet had been only a little farther from the sun, it is possible that Earth might have been locked in a *permanent* ice age. As on Mars, any water on planet Earth might have been frozen into polar ice caps or been stored as permafrost in the soil. Neither scenario sounds particularly appealing.

Of course, no one can speak with certainty about these things. Life, especially, has a way of resourcefully altering the planet's surface and atmosphere to suit its own purpose. But this much has been made clear by the record of the rocks: On several occasions during the Earth's 4.6 billion years, great areas of the continental crust have been covered with ice. Ordovician rocks in the North African Sahara Desert were scratched and grooved by ice when that continent lay nearer to the South Pole. The extent to which the continents have been covered with ice depends upon a delicate budget of energy from the sun, and how that energy is distributed across the planet.

How much of the sun's energy is retained on the surface of Earth, and how much is reflected back to space, depends upon the Earth's average albedo, or reflectivity. Deserts, cloud cover, forests, water, and ice all have different albedos. Surfaces covered by ice, like Greenland or Antarctica, are the most reflective. Water has a low albedo, and absorbs a larger fraction of the sun's rays than land or ice. When sea levels are high, as when rifting is especially active along midocean ridges (the ridges displace water), more of the continental margins will be covered by sea and more of the sun's heat will be absorbed. The warm sea will heat the atmosphere above it, and currents will carry the heat toward the poles. And so the tempo of tectonics will affect the Earth's supply of surface heat.

The rearrangements among continents and growth of mountain ranges also affect the way in which currents in the oceans and atmosphere distribute energy. When continents are near the poles, as today Antarctica is, ice sheets are likely to grow. Because ice is so highly reflective, an ice sheet's growth is self-reinforcing. As an ice sheet grows, it will reflect energy from the sun back to space. The Earth will cool further, and so the ice expands further. The planet is like a delicately tuned, electronically controlled mechanism, with dozens of interconnected circuits, laced with feedback devices. The device is so complex that no one yet knows exactly why the great ice ages have come and gone.

But what of the most recent ice age, the one that began only a few million years ago? Surely so recent a puzzle must be easy to solve. The trigger that initiated the current succession of ice ages is actively debated by geologists and climatologists. Some scientists believe that creation of the isthmus at Panama was the trigger for the recent ice ages on the northern continents. The isthmus is relatively new. Beginning about twenty million years ago the Earth's crust between North and South America was caught in a squeeze of plates. The Pacific Ocean floor was pushed down below the floor of the Caribbean Sea. Above the subducting plate the sea floor was crumpled upward and injected with volcanic lava. By three million years ago, a solid land link had been established between the two continents. Before the link was complete, a part of the warm-water North Atlantic Equatorial Current passed through the Caribbean and into the Pacific, through the gap between North and South America. When the isthmus was at last fixed, the current was deflected northward, through the Gulf of Mexico, and reinforced the Gulf Stream.

The Gulf Stream is the ocean current that brings warm water to the North Atlantic. Two conditions are necessary for an ice age to begin—cool continental temperatures, and sufficient moisture in the air to fall as snow. A strengthened Gulf Stream may have contributed to the second requirement by bringing warm water to the North Atlantic where it could evaporate into the air. Winds then carried the moisture-laden air over the continents, where it fell as snow. An ice sheet then began to grow. The slip and slide of crustal plates far to the south, which created the isthmus at Panama, may have been the factor that set glaciers moving across the Northeast landscape!

Another possible cause setting off the ice ages may have been the Himalayan uplift as India collided with Asia. This connection may seem improbable considering how far away those mountains are from the place where the ice sheets grew. Yet the Himalayas, which have gained much of their lofty height in the past few million years, appear to profoundly influence atmospheric circulation, especially the "jet stream" wind current in the upper atmosphere. The mountains appear to cause the large meanders, or wavy loops, seen in the jet stream on a typical weather map. By causing the jet stream to dip southward over North America, the Himalayas have contributed to colder temperatures on our continent. Cold continental temperatures fit the other requirement for an ice sheet's growth.

Pulsing Glaciers

Stand in the glacier-carved bowl of Tuckerman's Ravine, on the eastern flank of New Hampshire's Mount Washington, even as late as midsummer day, and you can feel the chill of the ice age. This shadowy cavity on the mountainside may have been one of the last places in the Northeast where the ice age lingered. Even today conditions in Tuckerman's Ravine are not far from those in which a mountain glacier will grow. At this last place in the Northeast for late-season skiing, patches of snow endure until June or July. The first snows of the following winter come a few months later. With only a small drop in average temperature, snow would accumulate from year to year, and once again a glacier would build in the ravine and send its weighty, sharp-clawed paw reaching down the mountain.

Three million years ago snow began to accumulate from year to year in central Canada and in parts of northern Europe and Asia. The weight of the ice squeezed the glacier out along its edges. This increased area caused further cooling, by reflecting sunlight to space. The glacier expanded further. It continued to grow until it reached the sea—where the unsupported ice broke off into bergs that floated away—or until it had pushed far enough south so that the ice melted as fast as it arrived from the north. In North America, the glaciers reached as far south as the present Ohio and Missouri river valleys.

For three million years the northern ice sheets have expanded and contracted rhythmically. Again and again they have extended their sway across the continents. Again and again they have shrunk back to small enclaves in northern Canada and the mountains in the West. Small periodic wobbles and variations in Earth's orbit change the amount of radiant energy the planet receives from the sun. The most significant variation has a period of 100,000 years. Once every 100,000 years the Earth's climate has warmed sufficiently to allow the glaciers to recede. We are in one of those periods of remission now. Less ice lies on the northern continents today than at any time since the last interglacial period 100,000 years ago. North America is basking today in an infrequent interlude of warmth.

One hundred thousand years ago the most recent ice sheet began to grow. Year by year the glacier expanded until all Canada was covered. By then the Northeast climate had cooled to tundra conditions. The forests fell away under the chill, and the ground was hard with frost. Slowly, by inches or feet per year, the glaciers advanced, pushing into northern New England and down across the Adirondacks. Great tongues of ice probed the northern mountain valleys, curling around the highest peaks, then coalescing, until only the tops of the highest mountains stood like rocky snow-covered islands in a sea of ice. The southern glacial margin was a wall of ice that reached from horizon to horizon.

For thousands of years the Northeast lay beneath the glacier. But the glacier was not a static pile of ice, any more than the glaciers that now cover Greenland and Antarctica are static. The ice was in motion, moving from the centers of accumulation out toward the edges. And it was dirty ice, filled with the rubble it scraped up from the land. It moved across the bedrock like a giant rasp, ripping and tearing. V-shaped valleys in the Adirondacks and the New England mountains, sliced deep by water, were rounded and smoothed by ice. Stand today at the head of Crawford Notch in the White Mountains and look down toward the foot of the valley: The valley walls are U-shaped—the only contour consistent with a tongue of moving ice, which unlike water is not free to follow the lowest creases in the land. The same is true of Franconia Notch, Evans Notch, Pinkham Notch, and the other notches in the New England mountains. Rushing water flowing on the floor of these ice-rounded valleys is today working to restore them again to a more precipitous shape, but for the time being they show ice's rounding influence.

If you could scrape the Northeast bedrock bare—shave away the forests and the farms, the soil and the loose rocks, down to the bony, solid crust of the Earth—then the thousands of years of grinding by ice would be patently obvious. In many places the bedrock is as smooth as a planed board, and everywhere scratches and striations mark the direction in which the ice moved. Once-symmetrical outcroppings are now steep-sloped and shattered on their southern side—the down-glacier side—where ice in its passage plucked away the rock. Valleys are smoothed as if scooped by a trowel. And on the highest peaks, such as New Hampshire's Mount Washington and Maine's Mount Katahdin, peaks that stood above the ice, the glaciers that nested in the highest valleys have gouged great semicircular bowls of rock called cirques. Tuckerman's Ravine is a cirque; it looks as if it had been cut by an ice-cream scoop. On Katahdin's south side, ice attacked the mountain so vigorously from opposite sides that nothing is left of the original ridge but a precipitous "Knife Edge," along which a terrifying and spectacular walking trail will take you today.

The Ice Recedes

Thirteen thousand years ago the Northeast climate began to warm. Melting accelerated at the margins of the continental ice sheets. Slowly the ice edge began to recede, northward. Continental glaciers move in one direction only, outward from centers of accumulation. They do not move backward. When a glacier retreats, it *melts* backward, even as it moves falteringly forward. The glacier edge is determined by a balance between forward motion and backward melting. When the climate began to warm 13,000 years ago, increased melting began to pare the glacier back.

And as the glaciers receded, the Northeast was given its final form. The works of a billion years—the mountains, the rift valleys, and the folded rocks, the residue of the crash and drift of continents—were now touched up and decorated by receding ice. And perhaps the most prominent modifications left behind by retreating ice were the great southern moraines.

When a bulldozer moves forward, it pushes all before it. When it backs up, it leaves a pile of debris at the place marking its farthest advance. As the continent-spanning glacier grew south from Canada, it pushed before it some of the matter that lay in its way. During the long time that the glacier lay upon the Northeast, it eroded the ground under it. Part of the eroded material was carried by the moving ice to the southern margin, and there—as the ice melted—the material was dropped. Meltwater streams flowing away from the glacial margin carried off fine silts and sands. Material too heavy or coarse to be carried by water was dumped at the glacier's foot, eventually building a lofty ridge—two ridges, actually, for the southern terminus of the glacier resided for a long time at two places (Figure 9.2).

The first moraine built by the glacier stands above the restored sea at Nantucket, Martha's Vineyard, Block Island, and Long Island (the southern arm of the fork at the eastern end of the island). Cape Cod, the Elizabeth Islands, Watch Hill in southern Rhode Island, Fishers Island, and Long Island's north fork are parts of the second moraine. These islands and peninsulas are among the youngest landscape features in the Northeast. They are scrap piles, heaps of rubble scraped from the land to the north. In them you might find rocks carried by ice all the way from Canada. Pulverized grains of granite scraped from Mount Washington are among the sands on Cape Cod. When Army Engineers dug the Cape Cod Canal earlier in this century they heaved and blasted their way through half of New England—giant boulders, middle-sized boulders, small boulders, all locked in a matrix of erosional debris, scooped or plucked from the terrain to the north and carried south by ice. The New York Harbor Narrows are a river-cut gash in the same ice-constructed ridge, now spanned by one of the world's great bridges—the Verrazano. At New York Harbor, the moraine

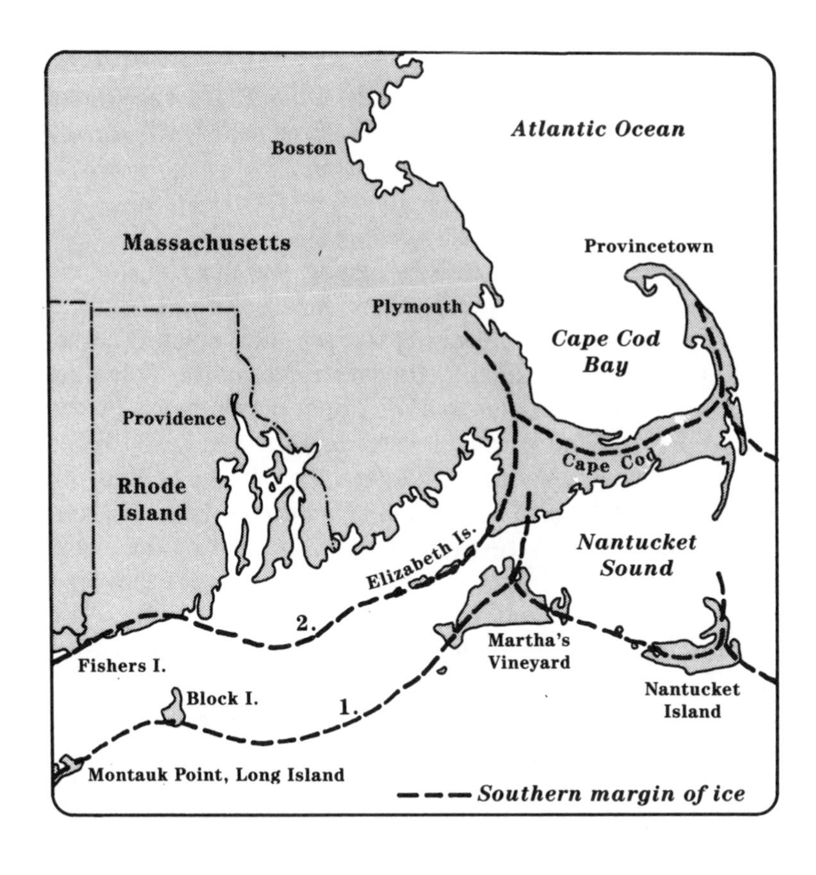

FIGURE 9.2—At two distinct times, the southern margin of the ice sheet remained stationary long enough to build up mighty ridges of erosional debris called *moraines*.

(here one thick ridge) turns south to buttress the Staten Island shore, and then bends at an elbow near Perth Amboy to arch up across northern New Jersey.

Cold Lakes

As the ice fell back in New Jersey, the moraines, along with more ancient hills and ridges, acted as dams, holding back meltwaters from reaching the sea, and creating lakes. The natural basin formed by the lava ridges of the Watchung Mountains and the Ramapo Mountains, now drained by the Passaic River, filled with meltwater to form cold Lake Passaic (Figure 9.3). A second and larger lake formed behind the southernmost loop of the terminal moraine—Lake Hackensack. The lake filled the present basin drained by the Hackensack River. Other freshwater glacial lakes covered Manhattan Island and parts of Long Island Sound. The swamps, marshes, and meadows in New Jersey and New York City are the last residue from these lakes. The Great Piece Meadows and Hatfield Swamp near West Caldwell, New Jersey, the Troy and Black Meadows near Whippany, and the Duck and Great Swamp near Chatham, stand on the clay-filled bottom of glacial Lake Passaic. The Hackensack Meadowlands are the floor of Lake Hackensack, and the Flushing Meadows in Queens are part of the floor of glacial Lake Flushing. If 13,000 years ago you had stood on the high Palisades ridge opposite Manhattan, you would have found yourself on the crest of a long, narrow island. To the north is the glacier edge, falling away toward Canada, a receding wall of ice stretching from east to west as far as the eye can see. On opposite sides of your island are broad lakes filled with frigid water that leaks away from the melting ice. The meltwater outflow is here temporarily kept from flowing out across the still-dry continental shelf to the distant sea by ridges of resistant Triassic lava and heaps of glacial rubble.

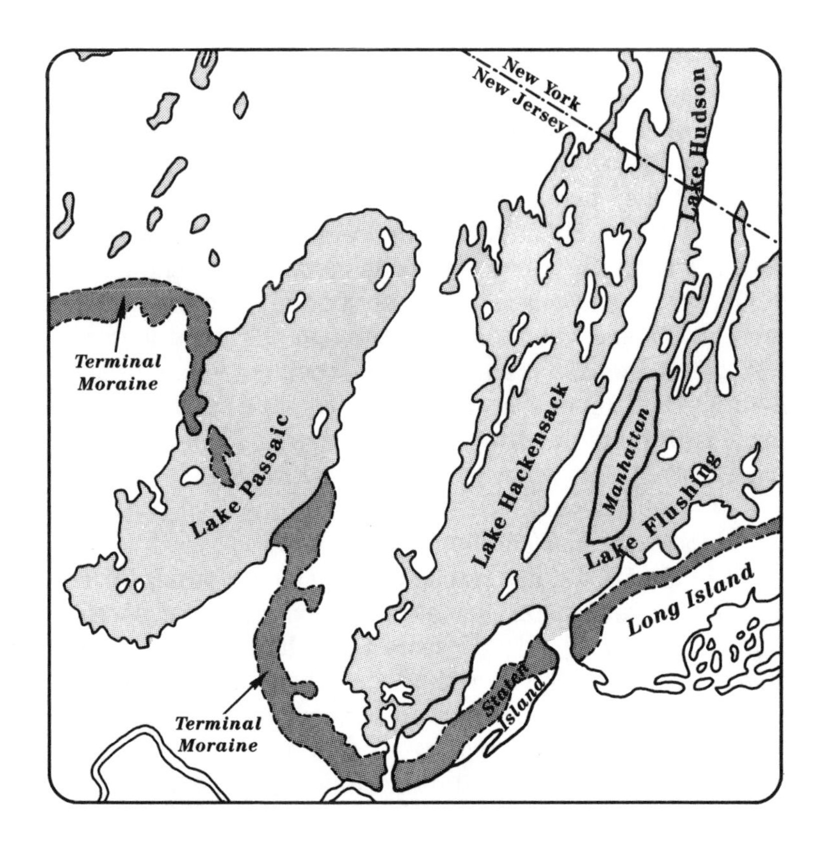

FIGURE 9.3—The location of these glacial lakes at the end of the last ice age, about 13,000 years ago, is based on a map provided by the New Jersey Bureau of Geology and Topography.

Other temporary glacial lakes were left behind by melting ice all across the Northeast. Among the most extensive was Lake Hitchcock, a 160-mile-long lake that filled the Connecticut Valley behind a morainal dam at Rocky Hill, seven miles north of Middletown, Connecticut. Lake Hitchcock lasted several thousand years. It came to an end about 10,000 years ago when its outlet stream, precursor of the present Connecticut River, cut through the dam at Rocky Hill. When the lake drained, winds moving across its sandy floor piled up dunes that can still be seen at Amherst, Chicopee, and Longmeadow, Massachusetts. Dune building ceased when vegetation reclaimed the lake floor.

Drumlins, Kettle Holes, and Eskers

The glacier did not recede in one smooth sweep. It fell back by fits and starts, sometimes stopping or temporarily edging forward. By 11,000 years ago it had released its icy hold on the land. In its retreat the ice left behind several geographic features that now add charm and variety to the landscape of the Northeast.

The oval hills of the Boston area—Bunker Hill, Breeds Hill, Dorchester Heights, and the Boston Harbor islands—are glacial *drumlins,* masses of clayey and gravelly material gathered and deposited under the ice in a way not yet fully understood by geologists but perhaps a little like a child rolling modeling clay beneath her hand. The sinuous Maine ridges, some of which run tens of miles north to south like abandoned railroad embankments, are *eskers.* They formed from stream tunnels under the ice that became filled with gravel and cobbles; when the glacier melted, the filled tunnels remained as winding ridges. The Cape Cod ponds are *kettle holes,* as is Thoreau's Walden Pond in Concord, Massachusetts; the ponds are the sites where blocks of stagnant ice broke off the retreating glacier; meltwater, carrying sand away from the glacier's edge, built up a plain of outwash material around the blocks; when the ice blocks finally melted they left depressions in the sand. Drumlins, eskers, and kettle ponds can be found in many places in the Northeast.

But one need not go looking for moraines, drumlins, eskers, and kettle ponds to see the work done by ice. Glacial deposits are almost everywhere in the Northeast, providing a thin rocky cover for the bedrock that is so typical in the region. Glaciers are dirty. They scratch and claw as they move. They break and pluck, and they carry along whatever they remove from the land. When the great continental glaciers melted, the erosional debris that they carried was dropped at the place where the ice melted, covering the landscape with a blanket of rubble. In some places meltwater flowing from the glacier sorted the rubble by degree of fineness. Sand and silt were readily carried by water and deposited in outwash plains. But over much of the Northeast the ground is covered with till, an intransigent hodgepodge of boulders, rocks, gravel, sand, and silt. Till is not the kind of soil that is ideally suited for farming. The ubiquitous stone walls in the Northeast were not built only to mark boundaries of fields; they were also repositories for the boulders removed from fields in an almost endless attempt to create plowable domains. No wonder Northeast farmers picked up and moved west when less stony land opened up beyond the Appalachians. When the Erie Canal and later the railroads were built, cities in the Northeast had access to rich farmlands to the west, and no longer were farmers forced to coax crops from an ice age's dumped detritus. Stone walls that were laboriously built to clear and mark fields now wander through woodlands like the silent remnants of a lost civilization.

Rising Seas, Rebounding Crust

In 1848, workmen building a railroad at Charlotte, Vermont, uncovered the skeleton of a whale. This was not some extinct monster from the distant past, like the dinosaur bones found on western mountaintops. These were the bones of a whale such as those alive and swimming today in the Earth's seas, and the clays in which the bones were found are no older than 10,000 years. A whale! In Vermont? How is this mystery to be resolved?

During the ice age as much as one quarter of Earth's land area was covered by ice. The ice was two miles thick near the regions with greatest accumulation and a half-mile or so thick near the edges. Ocean water was piled up on the land in the form of ice, and the level of the world's seas was several hundred feet lower than at present. As the continental glaciers began to retreat, the seas rose. At first—as the ice pulled back only a little way from the place of its most southerly advance—Long Island and Cape Cod stood as ridges on dry land. Only with the full retreat of the glaciers and the consequent rise in sea level did those morainal ridges acquire their present status as an island and peninsula.

The huge weight of the ice lying upon the land depressed the Earth's crust. The present ice caps on Greenland and Antarctica have depressed the underlying continents by a thousand feet or more. The Earth is plastic and deforms under stress, but it is also solid. When the stress is removed, the crust rebounds to its former level, but slowly. The glaciers' retreat was quicker than the crustal rebound. When the ice melted the seas rose, and large areas of the continents were flooded that today are dry land. The area around Boston Harbor and vast stretches of the New Hampshire and Maine coast were temporarily flooded, until the slowly rebounding crust tipped the waters back into the ocean basin. Deep clay beds along the Atlantic coast of New Hampshire and Maine are the residue of the time when those areas were the Atlantic floor. The Maine coast is *still* rebounding, and still more of the water presently in the estuaries along that coast is destined to be shrugged away.

Waters rising over an ice-depressed crust made of the

present Saint Lawrence River a vast ocean estuary. An arm of the estuary reached down the Champlain Valley into northern Vermont and New York (Figure 10.2). Whales swam in this great inland sea—and died. And the clay on the floor of that now-vanished sea preserved the bones of at least one of these post–ice-age denizens of Vermont.

Throughout the Cenozoic era the Earth's climate has cooled, a process culminating in the ice ages of the last few million years. During their greatest advance, the glacial ice sheets were more than two miles thick at their center and reached as far south as Nantucket, Martha's Vineyard, and Long Island. These islands, as well as Cape Cod, were formed by material deposited at the ice sheets' melting edges. Ice also sculpted many other features of the Northeast's landscape, including drumlins, eskers, and kettle holes.

Chapter Ten

Going with the Grain

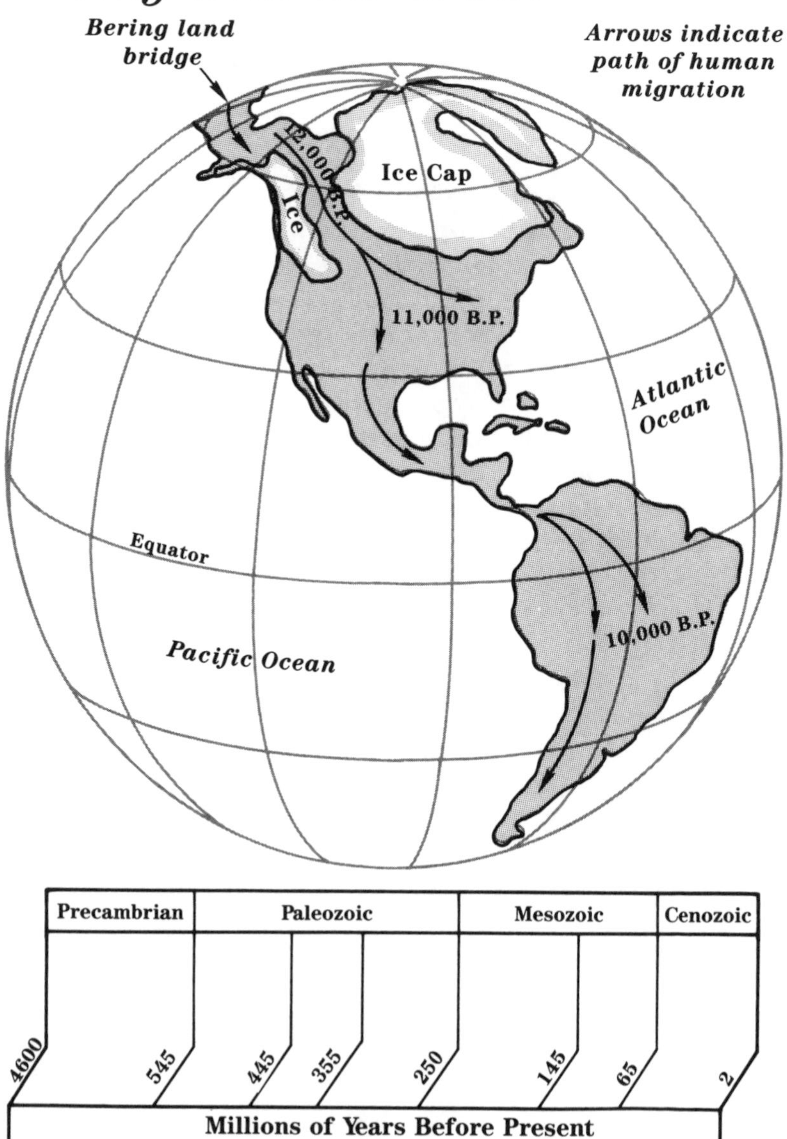

When the ice moved down across the Northeast from Canada it swept before it every shred of vegetation. Like an advancing bulldozer, the glacier scraped the landscape clean of fertile soil. Thirteen thousand years ago, when the climate began to moderate, the slowly retreating glacier laid bare a land devoid of life. In the lee of the ice, plants and animals began to reclaim the landscape.

Pollen, extracted from sediments in bogs in the Northeast and Canada, tells the story of the repopulation of the region. The first plants to reestablish themselves at the melting glacier's margin were tundra grasses and sedges and small plants such as wormwood, plantain, and fireweed. As the soil improved and the climate moderated further, the forest reappeared. First came fir and spruce, then white pine, hemlock, maple, beech, and birch. At last the landscape took on the mature forest appearance that greeted the Pilgrims when they stepped ashore at Plymouth.

But of course the Northeast had a human presence before the Europeans arrived and that story too is related to the ice. Only a few shreds of evidence suggest that human beings lived in the Americas before 12,000 years ago, and most of it is disputed. If people did reach the Americas before the end of the last ice age, they were few in number, left few artifacts, and had little effect on the environment. The first significant numbers of human beings seem to have arrived as the most recent ice age closed. They came from Asia, migrating across the Bering Strait.

The Bering Strait between Siberia and Alaska is only sixty-six miles wide. It is possible to see the glistening snow-capped peaks on one continent from the shores of the other. Human beings might have passed from one continent to the other by boat, or by walking across pack ice on the occasionally frozen strait. Their more likely means of crossing was dry land. At the peak of the ice age the sea level was lowered enough to create a land bridge between Siberia and Alaska. Surprisingly, even when the glaciers were at their maximum, much of Siberia and Alaska were ice free. Two conditions are necessary for a glacier to form—cold and moisture. Siberia and Alaska were cold enough, but the dry climate in those regions kept them unglaciated. Animal traffic moved both ways across the Bering land bridge. Asian species, such as the mammoth, the bison, and the saber-toothed tiger made their way to America. The American fox, ground squirrel, horse, and camel went the other way.

Sometime before about 20,000 years ago, nomadic tribes of human hunters entered Alaska from Asia following herds of grazing animals. Once in Alaska, they confronted an unbroken wall of ice blocking movement farther south across Canada. Only when the ice had begun to melt back far enough, about 12,000 years ago, did the bottled-up Alaskans have access to the south. An ice-free corridor opened up between the continental ice sheet in eastern Canada and the glaciers that mantled the western mountains. Through this ice-chilled corridor moved an eager wave of human migration. Within 2,000 years the wave had reached the southern tip of South America.

When those first bands of migratory hunters arrived in the temperate part of North America south of the ice sheets, they must have thought themselves in paradise. Here was game in plenty. The killing began, and with such vigor that many giant mammals in North America became extinct at this time—the glyptodont, rhino, camel, horse, saber-toothed tiger, giant ground sloth, mastodon, and woolly mammoth (Figure 10.1). Strong circumstantial evidence suggests that the agent of extinction was human overkill. Bones from butchered animals and carefully crafted flint spearheads have been found together at several sites. Whether the killing was for food, sport, ritual, or safety may never be known, but the killing spree set off an environmental overturn. In North America the "Age of Mammals" gave way to the "Age of Humans."

The slaughter was mitigated by beauty. The ice age

FIGURE 10.1—Extinct ice-age animals of North America, from top to bottom: giant ground sloth, saber-toothed tiger, camel, glyptodont

rigors evoked in the human mind a wonderful flowering of imagination. At the height of the glaciations we find in the caves of southern Europe magnificent works of art that have seldom been surpassed—carvings in bone and clay, and paintings of animals. In the cave inhabitants' decorated artifacts we find evidence for early religion, science, and technology. With intelligence and imagination came access to deadly force. People learned to use fire and flint in the struggle for survival. No longer would nature require millions of years to perfect tooth, claw, scale, or feather. No longer would size, swiftness, or armor ensure survival. From the end of the ice age onward, the scale and pace of environmental change would be shaped by human craft. When the Pilgrims arrived on these shores, they found a landscape already marked by human intervention. And with them, the Pilgrims brought the scientific and technological revolution that was transforming the European environment on an unprecedented scale.

The Human Settlement

Radiocarbon dating of living sites in the Northeast suggests that human beings arrived here even as the ice was melting back from the land. Their entry into the region presumably came around the southern end of the Champlain Sea (Figure 10.2). The first native Americans in the Northeast may have fished in glacial Lake Hitchcock and the Champlain Sea. But they would also have relied greatly upon mammals as food resources. The extinctions that decimated North American mammal species between 12,000 and 11,000 years ago presumably affected the Northeast, although the paleontological record in our region is scanty. The woolly mammoth, mastodon, muskox, horse, tapir, ground sloth, dire wolf, and giant beaver all disappeared at about the same time as the last of the ice. The bison and the caribou endured and perhaps became the chief game for the Indian hunters. The extinctions may also have brought

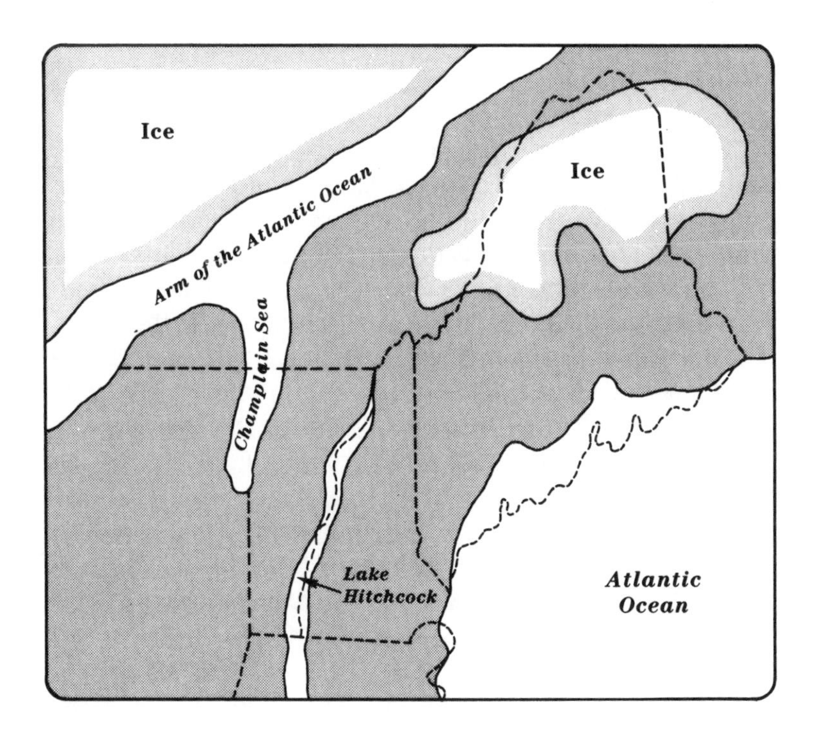

FIGURE 10.2—At the end of the ice age, the St. Lawrence River and Lake Champlain valleys were arms of the Atlantic Ocean.

about more reliance on plant gathering, in place of hunting, as a means of survival.

By the time the Pilgrims arrived in 1620, native Americans had established a characteristic pattern of life, in balance with the region's natural resources. The semipermanent village was the main unit of habitation. Indians cultivated corn and beans. They hunted and fished. They adapted their life to seasonal variations in the abundance of fish and game. They periodically burned small areas of forest to clear land for crops. During the 10,000 years that followed the mass extinctions, the native Americans in the Northeast left a relatively minor imprint on the landscape.

By contrast, only a few hundred years of habitation by Europeans utterly transformed the environment. The pattern of European colonization was controlled by geology: the Northeast is a crumpled and torn terrain. The crumples and the tears run mostly north to south, parallel to the lines of continental collision and rifting, and along these folds, creases, and rents in Earth's crust the present rivers flow. The Hudson River–Lake Champlain Valley, the Connecticut River Valley, and the valley of New Hampshire's Merrimac River gave the new settlers access to the interior. Of these, the Connecticut River Valley is rather different, and has more in common with the Newark Basin in New Jersey than with the other river valleys. The Newark Basin and the Connecticut River Valley are rift valleys, where the Earth's crust was wrenched apart and downfaulted in Triassic times, and filled with sediment. At the end of the last ice age, meltwater lakes (Lake Hackensack, Lake Hitchcock) filled the basins, and silt collected on the lake floors. Today, the rift valleys are the only rich agricultural land in the Northeast—as the earliest settlers soon learned.

Except for the Triassic rift valleys, bedrock is never far below the ground surface anywhere in the Northeast. The typical cleared field has an outcrop of bedrock somewhere in it, and elsewhere the rocky skin of the Earth is just lightly covered with glacial rubble deposited at the end of the ice age as the ice melted. Farming the rolling countryside in the Northeast was never easy. A field was said to be as likely to yield a crop of stones as of agricultural produce. The omnipresent stone walls of New England are an enduring testament to the backbreaking work required to make a living on this stony ground.

But the early colonists in the Northeast did clear the land, first along the shoreline, then in the relatively fertile river valleys, and, as good land became scarce, in the hilly uplands. Then, early in the nineteenth century, rich farmland was opened up for settlement west of the Allegheny Mountains. It made little sense for New Englanders to go on trying to coax food from stone. Farmers packed up their belongings and migrated westward, and those who stayed behind turned to manufacturing, an activity better suited to a land of rocky hills and fast-flowing rivers.

The rivers supplied the power that turned the wheels of industry. The raw materials of industry were mostly agricultural: textiles were a common product. The Northeast was never particularly rich in metals. The famed Saugus Iron Works, established by John Winthrop the Younger at Saugus, Massachusetts, in 1646, used bog and swamp sediments as the source of ore. ''Bog iron'' was not an adequate source of metal for industrial America; the Saugus Iron Works closed in 1684. Iron was also mined from the deformed Precambrian rocks of the old Grenville continent, in western Connecticut and the Precambrian New Jersey highlands, but those deposits also proved too slight to supply the modern appetite for iron and steel. Among other geologic resources in the Northeast are zinc deposits in the New Jersey Precambrian rocks and some small copper deposits in the central Connecticut valley. These copper ores were created at the time of the Triassic rifting, when hot, mineral-laden fluids from Earth's bowels were injected into the sedimentary strata of the rift basins. Copper was once profitably mined at Ely, Vermont, and Blue Hill, Maine, but these workings too were soon eclipsed by richer deposits in the West. Any metals nature gave the Northeast were given in stingy quantities.

The region is somewhat more generously endowed with building stone. Vermont's marbles, the metamorphosed carbonates deposited on the Precambrian Grenville shore, are justly famed. "Brownstones" in Connecticut's central valley, formations of oxidized sediments deposited in the downfaulted Triassic rift basins, were quarried from 1640. The stone was shipped by boat down the Connecticut River to the sea, and thence up and down the eastern seaboard. We still refer to buildings constructed with this material as brownstones. But the very quality that made brownstone so easy and economical to quarry—its softness— meant it was never a very durable building material. Handsome brownstone buildings in our eastern cities are slowly crumbling. The brownstone quarries closed in the 1950s, following the pattern set by most other quarrying operations in our region. Six hundred million years on a dynamic continental margin produced a landscape rich in natural beauty but scantily endowed of geologic resources of economic worth.

The Grain of the Land

The early settlers in the Northeast made do with the gifts that nature buried in the ground or scattered upon it, first by farming, then by manufacturing. By the end of the nineteenth century, the mill town had replaced the farming hamlet as the characteristic northeastern settlement. Now a new problem arose: How do you get agricultural produce from west of the Alleghenies to the population centers in the East, and how do you move manufactured goods in the opposite direction—against the geological grain of the land, those ancient north-south-trending crumples and folds? The Mohawk River across upper New York State was a unique natural channel of communication from the Great Lakes Region to the Hudson Valley, and soon it was exploited, first by the Erie Canal builders in 1825, and later by the railroads. At the junction of the Mohawk and Hudson Rivers, near Albany, this natural east–west passage ran up against an almost unbroken wall of mountains (along the line of the Taconic orogeny) that reached from New York City to Canada. In 1874 the Albany-to-Boston railroad breached the wall with the six-mile-long Hoosac Tunnel, through the narrowest ridge in the Berkshire Hills. In its time the Hoosac Tunnel was considered an engineering wonder of the world. It was the symbol of a new power—the power to subdue geological hindrance.

Today, the Massachusetts Turnpike follows closely along the line of the Albany–Boston railroad but climbs up and over the uplands, slashing where necessary through solid rock, shaping the mountains to human design. Here, as elsewhere in the Northeast (Figure 10.3) where the interstate highway system runs against the grain of the land, the huge canyons of human-engineered road cuts have opened up for geologists, professional and amateur, wonderful views of the "inside" of Earth's crust. In the seams and folds of rock laid bare alongside the road, the modern traveler has vivid access to the story that is written in stone.

The technological know-how that enabled the human species, alone among all species of life on Earth, to go "against the grain" is a new term in the geological equation. Work that the ice-age glaciers required thousands of years to do, human beings can undo in a few years; as the Army Corps of Engineers did in moving a glacial moraine to build the Cape Cod Canal. Rivers are dammed, estuaries bridged, marshes drained, mountains moved. North of New Haven, tunnels carry a parkway through a wall of Triassic lava hundreds of feet thick. In Boston, part of the glacially depressed harbor was filled to make a place for Logan Airport.

This power to move mountains and fill bays includes the

FIGURE 10.3—A roadcut along a Connecticut highway exposes folded metamorphic rock.

power to change the environment on a global scale. Many scientists are concerned that carbon dioxide emitted into the atmosphere by burning fossil fuels, and the diminishment of nature's capacity to take carbon dioxide from the atmosphere caused by cutting tropic forests, are even now inexorably warming the Earth's climate. As the climate warms, ice will melt and sea levels rise. By early in the next century, beachfront property all along the Atlantic shore of the Northeast (and shores everywhere) may be threatened by rising water. Accumulation in the atmosphere of chemicals called fluorocarbons may be drastically depleting the upper-atmosphere ozone layer that protects life on Earth from the sun's ultraviolet radiation. Sulfur emissions from factories in the Midwest are carried by prevailing westerly winds to fall as acid rain into lakes and ponds in the northeastern wilderness, killing fish and collapsing food chains.

None of these environmental modifications are new in themselves. Sea levels have risen and fallen throughout the Earth's past as continents rifted and drifted across the face of the globe. Mountains have been pushed up by the engine of plate tectonics and brought down by wind, frost, and water. Respiring bacteria changed the atmosphere (by adding oxygen) in a way more drastic than anything human beings have done. What is new about the human factor in the geologic equation is the *time scale* of change—change that occurs faster than the capacity of evolution to adapt or the "balance of nature" to repair.

Change is inevitable. Human intervention in the environment is inevitable, even desirable. What is required is a measure of caution when our interventions go against the grain. The grain of the planet is marked in stone, in the stream of life, and in time. In this book we have traced the grain of one special landscape, that of the Northeast, across the geological eons and across the planet's dynamic face. Respect for the grain is a condition for our future well-being. Knowledge of the grain is the prerequisite for respect.

When the most recent ice sheet began retreating 12,000 years ago, human beings migrated to North America from Asia across the Bering Straits. The extinction at that time of many giant American mammals, including the saber-toothed tiger and the woolly mammoth, may have resulted from overhunting by those early Americans. When Europeans arrived, in the early 1600s, their agriculture and industry began transforming the northeastern landscape. Today, we are altering the environment on a global scale with too little thought about the possible consequences.

Bibliography

These are a few of our favorite books (and one map) about the landscape of the Northeast. All are written for the general reader. Some are out of print, but should be easy to find in libraries.

American Association of Petroleum Geologists. *Geological Highway Map of the Northeastern Region.* Available from the AAPG, P.O. Box 979, Tulsa, Oklahoma 74101. More than a geological map; a wonderful two-sided compendium of information on geological history, fossil and gem sites, and more.

Bain, George. *The Flow of Time in the Connecticut Valley.* 1963. Connecticut Valley Historical Museum, Springfield, Mass., and Pratt Museum of Amherst College, Amherst, Mass. A little classic. No mention of plate tectonics, of course, but the book is still delightful to read, and it will lead you to interesting sites in the Connecticut Valley.

Bell, Michael. *The Face of Connecticut: People, Geology, and the Land.* 1985. Bulletin 110 of the State Geological and Natural History Survey of Connecticut. ''Bulletin'' is a prosaic title for this handsome and engaging book. Well written and illustrated with fine photographs and drawings.

Chamberlain, Barbara Blau. 1964. *These Fragile Outposts: A Geological Look at Cape Cod, Martha's Vineyard, and Nantucket.* Natural History Press, Garden City, N.Y. A marvelous book. More than geology, more than human history: An intelligent and affectionate appraisal of a specific geography.

Cronon, William. 1983. *Changes in the Land: Indians, Colonists, and the Ecology of New England.* A Yale scholar considers the ways in which human beings (of two cultures) reacted to and modified the physical geography of New England.

Jorgensen, Neil. *A Guide to New England's Landscape.* 1977. Globe Pequot Press, Chester, Conn. The perfect companion to our book. Full of valuable suggestions on places to go and things to see. An indispensable addition to the amateur naturalist's library.

Redfern, Ron. *The Making of a Continent.* 1983. Times Books, New York, N.Y. A lavishly illustrated geological history of the entire continent, but with much material relevant to understanding our region. The basis for a television series of the same name.

Schuberth, Christopher J. 1968. *The Geology of New York City and Environs.* Natural History Press, Garden City, N.Y. Comprehensive and full of technical information, but completely accessible to the general reader. If you live in New York City, and if you enjoy our book, this is the one you will want to buy next.

Strahler, Arthur N. 1966. *A Geologist's View of Cape Cod.* Natural History Press, Garden City, N.Y. A classic of scientific exposition. Short, but jam-packed with information.

Thomson, Betty Flanders. 1977. *The Changing Face of New England.* A literate and loving journey through the natural and human landscapes of New England.

Van Diver, Bradford B. *Roadside Geology of New York.* 1985. Mountain Press Publishing Company, Missoula, Montana. Should be in the automobile glove compartment of every New Yorker who is interested in the landscape.

Van Diver, Bradford B. *Roadside Geology of Vermont and*

New Hampshire. 1987. Mountain Press Publishing Company, Missoula, Montana. Another of Van Diver's superb guides to the geology of our region.

Wolfe, Peter E. *Geology and Landscape of New Jersey.* 1977. Crane Russak, New York. Textbook-like in its completeness, and a useful guide to the landscape. We find this book to be an invaluable resource.

Index

Page numbers in italic refer to art and maps.

Index

2005-135
c.1 p/Amazon

Index